CLIMATE CHANGE AND GLOBAL WARMING

Edited by **Ata Amini**

Climate Change and Global Warming
http://dx.doi.org/10.5772/intechopen.74882
Edited by Ata Amini

Contributors

Juergen Pilz, Firdos Khan, Tauisi Taupo, Cornelia Van Wesenbeeck, Kelvin S. Rodolfo, A. Mahar F. Lagmay, Rodrigo C. Eco, Tatum Miko L. Herrero, Jerico E. Mendoza, Likha G. Minimo, Joy T. Santiago, Jenalyn Alconis-Ayco, Eric C. Colmenares, Jasmine J. Sabado, Ryanne Wayne Serrado, Jyri Hanski, Riitta Molarius, Jaana Keränen, María-Luz Rodríguez-Blanco, M.M. Taboada-Castro, R. Arias, M.T. Taboada-Castro, Zenebe Mekonnen, Bandana Pradhan, Puspa Sharma, Pushkar K Pradhan, Xinzhi Mu, Ata Amini

First published in London, United Kingdom, 2019 by IntechOpen
IntechOpen is the global imprint of INTECHOPEN LIMITED, registered in England and Wales, registration number: 11086078, The Shard, 25th floor, 32 London Bridge Street
London, SE19SG – United Kingdom
Printed in Croatia

British Library Cataloguing-in-Publication Data
A catalogue record for this book is available from the British Library

Additional hard copies can be obtained from orders@intechopen.com

Climate Change and Global Warming, Edited by Ata Amini
p. cm.
Print ISBN 978-1-78985-813-6
Online ISBN 978-1-78985-814-3

For EU product safety concerns:
IN TECH d.o.o., Prolaz Marije Krucifikse Kozulić 3, 51000 Rijeka, Croatia,
info@intechopen.com or visit our website at intechopen.com.

Meet the editor

Dr. Ata Amini is an Associate Professor of water resource engineering at the Kurdistan Agricultural and Natural Resources Research and Education Center, AREEO, Sanandaj, Iran. He completed his doctorate at the University Putra Malaysia in 2009. In recent years, he has been in charge of many research projects and supervised many masters and PhD students in the field of water engineering and environment aspects. He is also the author of five books and three book chapters. Dr. Ata Amini has devoted his academic career to study integrated water resource management, river engineering, sediment, scouring and watershed management. He has more than 20 years of teaching experience and has published many academic papers and final research reports.

Contents

Preface

Life on Earth strongly depends on climate. Societies have had to cope with or adapt to climate as the establishment of humanity. The decision-makers in the environmental sector should be able to understand the leading causes of historical climate changes and global warming and to evaluate the projections of such changes in the future. Apart from hydrology, climate knowledge involves diverse areas of sciences comprising biology, physics, geology, chemistry, and even sociology. Analysis of all the components and causes of climate change and global warming and the interactions between them is out of the scope of any book. I have thus chosen here to provide necessary information on the interaction between human activities and natural phenomena on climate changes and global warming. It is now generally accepted that social movements are changing the configuration of our ecosystem. However, misperceptions of the solutions are increasing. The current knowledge, presented by the Intergovernmental Panel on Climate Change of the United Nations (IPCC), reveals that we can overcome climate change and global warming. Most likely, further changes and negative influences are unavoidable; however, we can prevent the dominant impacts of climate change, so that life remains manageable. In recent years, the decision-makers' attention has shifted the balance between the likely influences of climate change, and technological advances, economic costs, and societal adaptations that are essential for mitigation. In fact, achieving this goal needs strong and decisive actions to be taken now.

This book, *Climate Change and Global Warming*, brings together the engineers, scientists, socialists, policy makers and specialists of the world to critically look at the various aspects of climate change, and it is an attempt to look at the facts. The overall purpose of the book is to introduce the concept of climate change and its effects within the context of sustainable development. The book starts with an introductory chapter, which presents the reasons and solutions for climate change and global warming. This chapter reports a case study to demonstrate the role of human activities on climate change and vice versa and the needs for an adaption to the changes that are occurring. This book is structured into two sections, which group different aspects of the addressed matter, consisting of 10 chapters. Section 1 discusses detecting and treating manmade climate change and global warming that we are facing. Section 2 deals with climate change as a development problem. The chapters in this section discuss the impacts of climate change and how societies can initiate adapting to them through beneficial methods, exploring activities underway at local levels, adaptation decisions, and how the world can become better prepared to make adaptation choices.

This book could not have been written without the valued work of all the authors who put such outstanding manuscripts together. The editor would like to thank all the authors for their efforts and excellent collaboration in the preparation and editing of the chapters. Furthermore, I wish to appreciate greatly Ms. Sara Debeuc for her editorial assistance in the preparation of the chapters. It is hoped that the contents of this book lead to improvements in our knowledge and thoughts on the comprehensive management of climate change and global warming.

Dr. Ata Amini
Associate Prof.
Kurdistan Agricultural and Natural Resources Research and Education Center
AREEO Sanandaj, Iran

Section 1

Modeling and Detecting

Chapter 1

Introductory Chapter: Lake Urmia - A Witness to the Simultaneous Effects of Human Activities, Climate Change, and Global Warming

Ata Amini

Additional information is available at the end of the chapter

http://dx.doi.org/10.5772/intechopen.83605

1. Climate change and global warming

In spite of the irregular behavior on a daily basis, observations show that, in the atmospheric character of some geographic area, there is some long-time regularity. However, in the twentieth century, the world temperature increased which influenced the world climate. Climate scientists reported that humankind's growing emission of greenhouse gases will make the long-term change in the climate and global temperature. As organizations, governments, and people have moved onward with strategies and actions to decrease greenhouse gas (GHG) emissions, to adapt to the impacts of climate change, the need for information to support climate-related decisions has grown rapidly in recent years. There is a lack of credible information system to inform climate adoptions and estimate their efficiency.

Greenhouse gases enter the atmosphere through burning coal, natural gas, oil, solid waste, wood products, and trees. The greenhouse gases are emitted during the production and transport of fossil fuels or as a result of livestock and other agricultural practices. Changes in greenhouse gas emissions are influenced by many long- and short-term factors. Larger releases of greenhouse gases result in higher concentrations in the atmosphere. However, as a part of the biological carbon cycle via a certain chemical reactions, carbon dioxide, as a main greenhouse gas, is removed from the atmosphere when it is absorbed by plants. The removal and emissions of greenhouse gases by natural processes tend to balance. Human activities, since the industrial revolution, by adding heat-trapping gases to the atmosphere, have contributed substantively to climate change. It seems that there are enough scientific evidences to prove that the greenhouse gases caused the climate alteration and global warming [11]. In addition, we realized that global temperature was increased over the last decades. However as stated by [6], the climate change holds significant risks to the natural resources, environments, and societies on which they depend. Climate change and global warming affect human life, meanwhile human activities are also influencing climate.

2. The climate effects on human life

Humans for food, livelihood, commerce, natural resources, and security rely on earth which is a complex and dynamic system. Climate always influences the humans. In recent decades, regardless of the wealth and technology of modern industrial societies, in many ways, climate still affects human life. The impacts of climate change and consequently the rising greenhouse gas levels including warming temperatures lead to extreme weather events, altered weather patterns, crop failure, rising sea levels, ocean acidification, marine ecosystem shift, and homelessness. These impacts directly or indirectly threaten our health by affecting the weather we experience, the water we drink, the food we eat, the energy we use, the air we breathe, and other aspects of human health and well-being.

3. The human contribution to climate change and global warming

Although there are scientists who suspect that the human activities could have emotional impact on recent changes in climate and global temperature, the Intergovernmental Panel on Climate Change, IPCC, approved that the warming detected over last 50 years is attributable to man-made activities [7]. It seems that there is no doubt that human activity has changed the composition of world's atmosphere and environment. The noticeable problem is that human activities' influences will continue to alter atmospheric composition through the current century and hydrologic cycle and temperature are expected to be change inconsistently. Climate change and global warming, in complex ways, interact with other ongoing changes in human and environmental systems. For example, humanity's choices about land use, energy, and food makings affect and are affected by climate change and global warming. Although the details of how the upcoming influences of climate alteration will unfold are not as well understood as the basic reasons and mechanisms of climate alteration, we can rationally assume that the consequences of climate alteration will be more severe if actions are not taken to limit its magnitude and adapt to its impacts.

4. Climate change in Euphrates-Tigris Basin

The increase of the frequency and intensity of droughts in Asia and Africa is reported as an evidence of climate change in recent years [8]. Most regions in the Middle East including Iran are in the midst of a water crisis and their worst drought in decades. As a result, the water resources declined considerably. At the current rate of decline, the basin's water supply will not be enough to avert a widespread humanitarian crisis [2]. [10] used Gravity Recovery and Climate Experiment (GRACE) satellite mission and concluded an alarming rate of decrease in total water storage equal to 143.6 km^3 from January 2003 to December 2009. Amini analyzed the climate data from Iran and its riparian countries [1]. The results revealed the considerable spatial and temporal changes in the precipitation total series of regions during the last 10 years. The human activities such as dam construction and interbasin water transfer intensify such changes [1]. The most well-known evidence to demonstrate the effects of climate change and human activities is shrinking and drying up of Lake Urmia in Iran.

5. The Lake Urmia

So far managing renewable resources and infrastructure in much of the nation's experience is based on the historic record of stable climate. The Lake Urmia basin in northwest Iran is a closed drainage basin with an area of 51,876 km² and maximum depth of 16 m, making it one of the largest lakes in the world. In addition, Lake Urmia is the second largest saltwater lake in the world. As a result of arid to semiarid climate of the Lake Urmia basins, the agriculture is mainly dependent on irrigation. The sprinkler irrigation was developed, and large numbers of wells were excavated to supply the required water in the agricultural sector. The reduction in rainfall along with decreasing groundwater tables and a rising population in the basin will likely exert growing pressure to continue diverting stream flow before it reaches Lake Urmia. Over the last two decades, it was sharply dried and its surface reduced about 80% from 2003 [5]. The water table of Lake Urmia was decreased severely in recent years which caused serious socio-environmental consequences [4]. The dried lake bed is with a cover of salt, mainly sodium chloride, making a vast salty desert. The examinations for estimating the relative contributions of human activities as water resources development including agricultural development and dam constructions and climate change have been conducted by [9]. The results showed that the variability in inflow as a result of human activities is more prominent than the variability in precipitation. In addition, the flow origin from the Lake Urmia was decrease up to 40% due to the development of irrigated area which increased pressure on the water availability in the watershed. Shadkam showed that over 30 years annual inflow to Lake Urmia has thrown down by 48% which 3/5 and 2/5 of this change was caused by climate change and water resource development, respectively, as misconduct in human activities [9]. **Figure 1** shows the changes in water level and water surface in Lake Urmia within the last two decades.

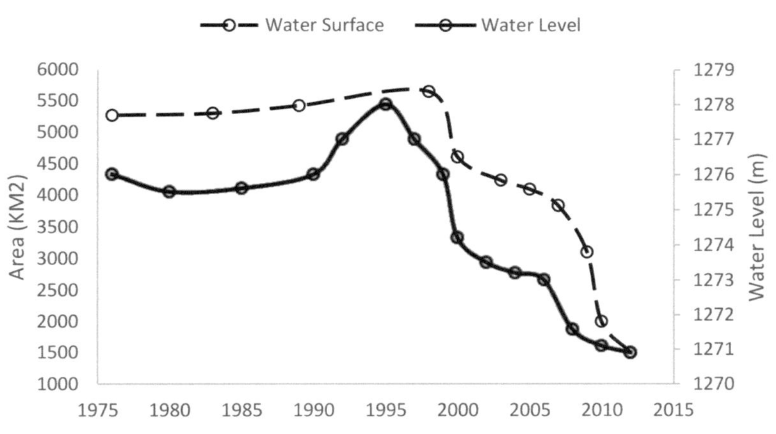

Figure 1. The changes in surface and water level of Lake Urmia [4].

Amini and Hesami got the same results for two main subwatersheds of Urmia basin. They reported that the total irrigated area increased by ~20% between 1989 and 2000 and the net irrigation water requirement, NIWR, for a crop grown in the region in 2000 was slightly higher than that in 1989 as a result of global warming [4].

Hosseini-Moghari using spatial and hydrology and in situ data assessed the contribution of human acts to the decline of inflow water for the lake, the groundwater table reduction, and water storage in the lake reservoir [5]. They found that 50% of the total basin water loss is due to sectorial management of water resources in the region. The human interaction with the lake environment in Lake Urmia basin caused about 8 BCM of groundwater losses within 11 years. However, the climate change and drought are the causes for 40% of lake shrinking in the last two decades. Such findings may be used to support decision-maker in water sector to restore the Lake Urmia. It seems that there is no debate that the Lake Urmia diminishing was caused by a combination of water withdrawals for irrigation and climate change and global

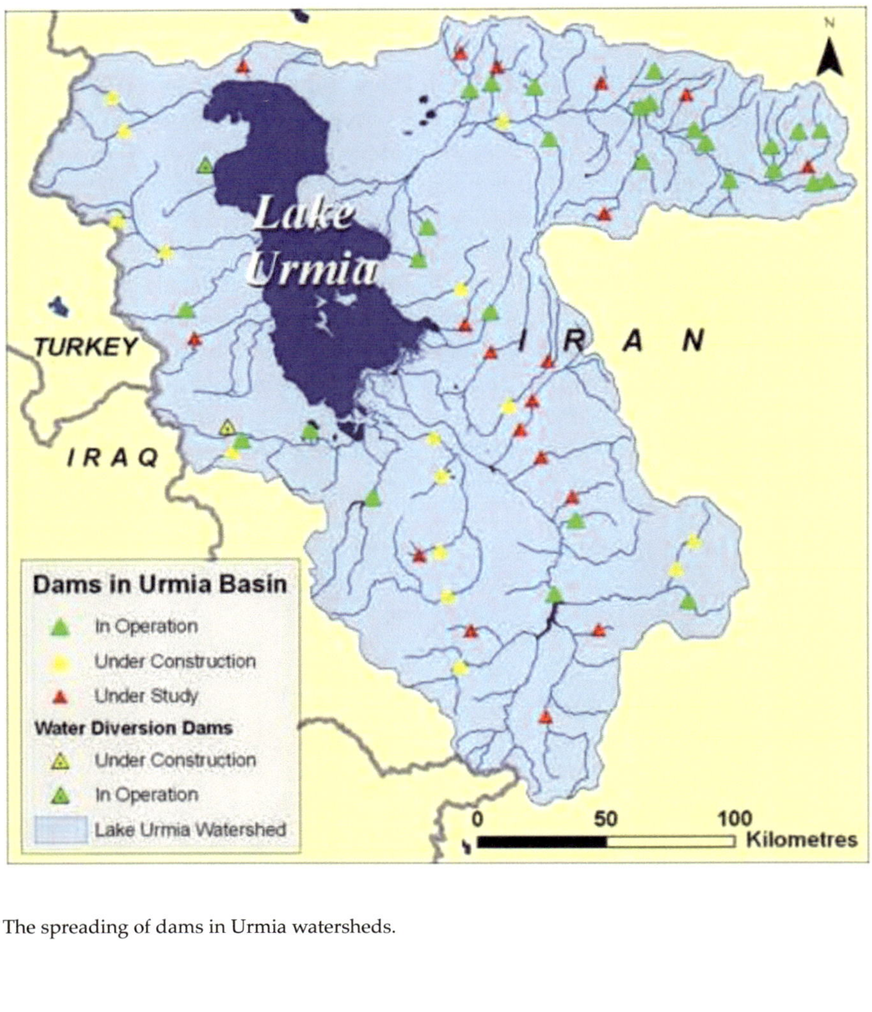

Figure 2. The spreading of dams in Urmia watersheds.

warming [3]. The lake restoration was taken placed by local, national, and international organizations. However, the decision-makers inappropriately put the man-made infrastructures such as dams and interbasin water transfer, which are unsustainable projects in water sector, to increase the flow toward the lake. In fact, construction of numerous dams in the watershed has choked off water source from the mountains towering on either side of the Lake Urmia. **Figure 2** shows the existing, under construction, and under study dams in the basin of Lake Urmia. **Figure 2** shows intensification in diversion of surface water and water utilization further than current levels which at present appear to be unsustainable.

Reducing the basin's water consumption and demand-side plans must be considered immediately by the official organization. Moreover, the current projects for water transfer need drastic revision. Such projects have had harmful socioeconomic and environmental side effects in other parts of the basin. To maximize the welfares of the water resources, an integrated management of the water resources and agricultural development based on ecological potential of the basin including changing the rules of water pricing and recognition of cost allocation outlines as a prospect is required.

6. Need for adaptive management

The IPCC clearly confirmed in its Fifth Assessment Report that the climate change is existent and its primary cause is human-made especially through burning fossil fuels and other activities that release heat-trapping greenhouse gases into the atmosphere. There is no debate that the human needs to take actions to adapt and lessen the impact of climate change. In fact, there is a hierarchy of governor plans that can assist to protect population healthiness and welfare. The predictions of future climate change and global warming and defined scenarios by IPCC indicate that unless substantial and continued activities are taken to decline emissions of GHGs, world will continue to warm. Thus, we can reasonably expect that such changes are driving several related and interacting deviations in the environmental system and consequently the human life. In the case of Lake Urmia, a national committee in order to lake restoration was formed in 2013 called Lake Urmia Restoration National Committee (ULRNC). The ULRNC presented a road map and action plan based on integrated water resources/watershed management in cooperation with the universities in Iran and international organizations. The ULRNC reviewed 19 proposed programs and approved a 10-year program in three phases such as stabilizing the present status, restoration, and sustainable restoration. Although the provincial office claims that the restoration continues as scheduled, due to drying up the found and misconducting of management in water and agricultural sectors in the region, it seems that Lake Urmia will most likely fail to attain an ecological balance within a 10-year program by 2023. Although the specifics of how the upcoming impacts of Lake Urmia decline will unfold are not as well recognized as the main causes and mechanisms of lake shrinking, it is rationally expected that the avoidable consequences of lake drying will be more severe if activities are not taken to limit its level and adapt to its influences.

Author details

Ata Amini

Address all correspondence to: ata_amini@yahoo.com

Kurdistan Agricultural and Natural Resources Research and Education Center, AREEO, Iran

References

[1] Amini A, Zareie S, Taheri P, Wan KBY, Mustafa MR. In: Bucur D, editor. Drought Analysis and Water Resources Management Inspection in Euphrates—Tigris Basin, River Basin Management. Romania: InTech Publication; 2016

[2] Chenoweth J, Hadjinicolaou P, Bruggeman A, Lelieveld J, Levin Z, Lange MA, et al. Impact of climate change on the water resources of the Eastern Mediterranean and Middle East region: Modeled 21st century changes and implications. Water Resources Research. 2011;**47**:W06506

[3] Hassanzadeh E, Zarghami M, Hassanzadeh Y. Determining the main factors in declining the Lake Urmia level by using system dynamics modeling. Water Resources Management. 2011;**26**(1):129-145. DOI: 10.1007/s11269-011-9909-8

[4] Hesami A, Amini A. Changes in irrigated land and agricultural water use in the Lake Urmia basin. Lake and Reservoir Management. 2016;**32**(3):288-296

[5] Hosseini-Moghari SM, Araghinejad S, Tourian MJ, Ebrahimi K, Döll P. Quantifying the impacts of human water use and climate variations on recent drying of Lake Urmia basin: The value of different sets of spaceborne and in-situ data for calibrating a hydrological model. Hydrology and Earth System Sciences. 2018. DOI: 10.5194/hess-2018-318

[6] IPCC. Climate Change—The Physical Science Basis—Chapter 8. WGI Contribution to the Fourth Assessment Report of the Intergovernmental Panel on Climate Change. 2007. Available from: http://www.ipcc.ch

[7] IPCC. Climate Change: Impacts, Adaptation and Vulnerability. Contribution of Working Group II to the Third Assessment Report of the Intergovernmental Panel on Climate Change. NY, US: Cambridge Press; 2001

[8] McMichael AJ, Campbell-Lendrum DH, Corvalán CF, Ebi KL, Githeko A, Scheraga JD, et al. Climate Change and Human Health: Risks and Responses. Geneva: World Health Organization; 2003. p. 322

[9] Shadkam S, Ludwig F, Oel P, Kirmit Ç, Kabat P. Impacts of climate change and water resources development on the declining inflow into Iran's Lake Urmia. Journal of Great Lakes Research. 2016;**42**(5):942-952

[10] Voss KA, Famiglietti JS, Lo MH, Linage C d, Rodell M, Swenson SC. Groundwater depletion in the Middle East from GRACE with implications for transboundary water management in the Tigris-Euphrates-Western Iran region. Water Resources Research. 2013;**49**:904-914

[11] Wilson G, Fairén V, García-Sanz J, Zúñiga I, Otto D, Breitmeir H et al. Module 1: Introduction to climate change in the context of sustainable development, Climate Change: From science to lived experience, lifelong learning program. 2011. p. 180

Chapter 2

Assessing the Expected Impact of Climate Change on Nitrate Load in a Small Atlantic Agro-Forested Catchment

María Luz Rodríguez-Blanco,
María Mercedes Taboada-Castro, Ricardo Arias and
María Teresa Taboada-Castro

Additional information is available at the end of the chapter

http://dx.doi.org/10.5772/intechopen.80709

Abstract

Climate change is likely to have profound impacts on quality of water resources, by altering the magnitude and timing of nutrient delivery to stream network. However, water quality responses to climate change are difficult to predict, especially for nutrient loads because of combined uncertainties in water quality and quantity projections. In this study, the potential medium (2031–2060) and long-term (2069–2098) impacts of project changes in climate variables (temperature, rainfall and CO_2 concentration) on nitrate load in an Atlantic agro-forested catchment (NW Spain) were assessed using the soil and water assessment tool (SWAT) model. Climate change scenarios are based on data projected by regional models from the ENSEMBLES project and two CO_2 concentration scenarios. The results showed that nitrate load will increase in the future horizons (2031–2060, 6%; 2069–2098, 7%) in relation to current values (1981–2010), possibly due to the decline in grassland biomass, as well as an increase in the rate of mineralisation linked to the increase in temperature. Consequently, lower rates of fertilisers will be needed in these areas in future horizons, which should be taken into consideration when planning management strategies in order to mitigate the impacts of potential climate change.

Keywords: nitrate, climate change, SWAT model, agro-forested catchment

1. Introduction

Nitrogen (N) is a key nutrient in river systems. Its concentration and form in the aquatic environment generally reflect the integration of a number of sources within the catchment including point and/or nonpoint sources. In addition to these anthropogenic contributions, there are "natural" contributions from the mineralisation and nitrification of organic nitrogen in soils [1]. Within aquatic environments, nitrate (NO_3) is generally the dominant N fraction due to high NO_3 mobility in the environment [2], as a result of its persistence, high solubility and low reactivity. The N problem, due to the excessive application of fertilisers for crop production, has been of growing concern in recent decades as it affects the water quality for consumption, promotes the development of eutrophication and reduces biodiversity [3, 4].

Changes in the climatic conditions and, particularly, increases in air temperature, shifts in rainfall patterns and an increase in the frequency of flood and droughts alter the processes controlling the mobilisation and transfer of N from agricultural land to aquatic ecosystems [5–7]. Rising water temperatures influence biological processes and chemical characteristics of water resources, e.g. increasing the mineralisation of organic matter, decreasing oxygen solubility, increasing the variability in pH values and increasing growth rates of aquatic organisms. A lower water availability might affect the quality of surface water resources adversely and even have a significant negative impact on human health and the economic development of the entire region [7, 8].

Since protecting and restoring the aquatic ecosystem is a policy priority in Europe [9], catchment management planning should focus on adaptively managing climate change impacts, although climate change is not explicitly included in the European Water Framework Directive (WFD), because climate change is expected to have profound effects on water resources and water quality [10]. The facts show that effects of climate change have not been properly addressed in policy formulation and water resource management strategies in many regions around the world, probably due to the lack of accurate and reliable data on the possible effects of climate change on water resources. Nitrogen pollution is already considered as a global problem [11], and it is expected that N loss will aggravate in vulnerable areas due to climate change [5, 8]. For this, it is of upmost importance to assess and quantify the impacts of climate change and vulnerability of water resources and evaluate the efficiency of possible adaptation and mitigation policies.

While many researches in different study areas assessed the climate change impacts on catchment hydrology and water supply, water quality has been studied much less [6, 12–14]. So, little is known about the potential effects of climate change on biogeochemical processes at catchment scale and its associated impacts on water quality. So far, very few studies have addressed the water resource on the Atlantic region of Europe, and, consequently, little is known about the effects of climate change in water quality in this area. This can constitute a challenge because hydrological processes in this region will change in response to climate change [15–17], as they are expected to undergo a Mediterranization process. In fact, in the northwest of Iberian Peninsula, an increase in temperatures (2–3°C) is expected, particularly during spring and summer, with marked uneven distribution of rainfall, i.e. more rain in autumn but drier spring and summer [18]. Likewise, according to the Intergovernmental Panel on Climate Change [19], climate change will cause an increase in the future winter storm and flooding for the Atlantic region of Europe, which encompasses northern Iberian

Peninsula, western France, the Netherlands, Belgium, northern part of Germany, western Denmark and the United Kingdom. Uncertainties involved in climate predictions are larger in this transition zone of the Atlantic region than in other areas, e.g. the northern and the southern parts of the continent. Therefore, this transition zone should be an in-depth studied area.

This paper provides an overview of the effects of changes in climate on nitrate loads in NW Spain. Specifically, the main objective of this study was to assess the impacts of the potential changes in climate variables (temperature, rainfall and CO_2 concentration) on nitrate load in an Atlantic agro-forested headwater catchment located in NW Spain using the SWAT model (one of the most widely used catchment models throughout the world) in order to provide information to catchment management so that they can anticipate possibly the impact on water quality and design and implement the necessary mitigation actions within the catchment management programs.

2. Study area characterisation

The Corbeira catchment is located in the headwater of the Mero basin, at about 30 km from the city of A Coruña in NW SPain (**Figure 1**). The catchment covers an area of 16 km^2. Elevations range from 65 to 470 m a.s.l., with a mean slope of 19%. The geological substrate is

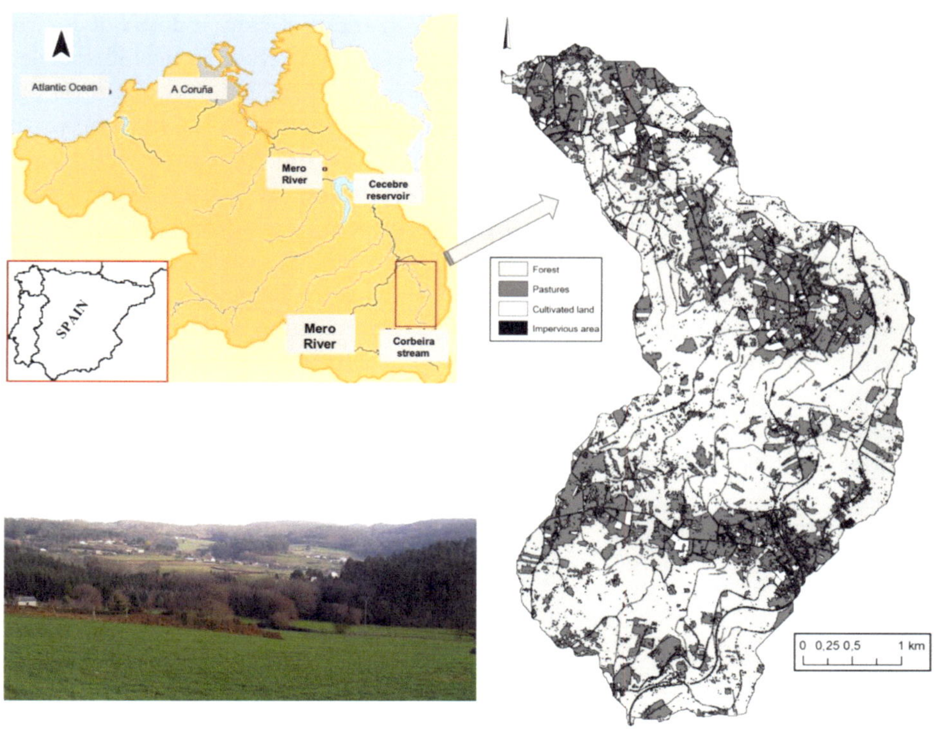

Figure 1. Location of the study area, general view of the Corbeira catchment and land-use map.

dominated by the basic schist of the Órdenes Complex [20], and the main soil types present in the catchment are umbrisol and cambisol [21], which represent 74 and 25%, respectively. Predominating land use is a forest covering 65% of the catchment area, followed by pasture (26%), impervious areas (5%) and cultivated land (4%). The population density in the catchment is low (35 inhabitants km^{-2}); there are no industries, and human activities are reduced to rural traditional agriculture and livestock (0.29 LU ha^{-1}).

The mean annual temperature is about 13°C, with mean annual minimum and maximum temperatures of 8.6 (February) and 18.4°C (July), respectively. The mean annual rainfall is about 1170 mm, more than 65% occurring between October and March. The annual mean flow rate is 0.18 m^3 s^1, and it is mainly supplied by groundwater. For more information of the study area, see Rodríguez-Blanco et al. [22–24].

3. Methodology

3.1. Model description

The SWAT model was developed by the Agricultural Research Service of the US Department of Agriculture (USDA) to quantify and predict the impact of agricultural management practices on water, sediment and agricultural chemical in large complex catchments [25, 26], although is has been satisfactorily applied in small catchments all over the world [10, 14–16]. It is a continuous, distributed model, although not completely distributed, since it does not use cells but divides the basin into sub-basins that are further divided into Hydrological Response Units (HRUs). For this reason, sometimes it is defined as semi-distributed. It is based on physical principles to describe the relationship between the input and output variables. It needs specific data from the catchment (climate, physical properties of the soil, topography, vegetation, soil management practice, etc.), which are used to model physical processes related to the movement of water and sediments, growth of crops and nutrient cycles. SWAT simulations can be separated into two components, the land phase for water and pollutants loadings to channels and the routing phase for in-stream water quantity and quality. Regarding nitrogen, the model simulates N transport and transformation at HRU scale; considering the processes of denitrification, volatilisation and organic N, stable organic N associated with humic substances and fresh organic N associated with the crop residues are distinguished. Nitrate can be transported from land to stream network via surface runoff, lateral flow and groundwater flow. A more detailed description of the SWAT model can be found in [25, 26]

3.2. Climate change scenarios

Reseach into the impact of climate in the future has focused on evaluating the effects that change in temperature, rainfall and CO_2 concentration might cause on nitrate load, following the methodology used in [15, 27]. Two simulation sets were used: one evaluated the response of the catchment to changes in single-climate variables (temperature, rainfall or CO_2 concentrations) and the other one assessed the impact caused by simultaneous changes in temperature, rainfall or CO_2 concentrations. In total, 14 different climate change scenarios were used (**Table 1**).

http://dx.doi.org/10.5772/intechopen.80709

Scenario	Modified parameter	Jan	Feb	Mar	Apr	May	Jun	Jul	Aug	Sep	Oct	Nov	Dec
1	T (°C)	1.0	0.9	0.7	0.9	1.1	1.0	1.3	1.5	1.4	1.2	1.0	1.4
2	T (°C)	1.6	1.5	2	1.7	1.9	2.1	2.5	2.7	2.4	2.8	1.6	2.3
3	T (°C)	1.8	1.6	1.6	1.7	2.2	2.5	2.6	3.0	2.7	2.5	2.2	2.2
4	(°C)	2.7	3.2	3.2	2.9	3.8	4.7	4.5	4.9	5.5	5.0	3.7	2.9
5	P (mm)	−1.2	1.0	−5.2	−12.7	−10.6	−11.4	−9.8	−11.8	−6.8	−8.2	−8.7	17.8
6	P (mm)	−26.0	−28.3	−12.9	−30.4	−15.5	−24.1	−17.2	−23.9	−30.7	−46.4	−37.9	54.7
7	P (mm)	−3.5	−1.0	−5.2	−17.6	−31.8	−21.0	−14.5	−16.4	−19.1	−27.4	−8.7	−8.9
8	P (mm)	36.6	17.2	−22.4	−26.5	−37.5	−30.2	−20.6	−28.5	−28.0	−56.1	−56.8	−29.6
9	CO_2 (ppm)	550	550	550	550	550	550	550	550	550	550	550	550
10	CO_2 (ppm)	660	660	660	660	660	660	660	660	660	660	660	660
11	T (°C)	1.0	0.9	0.7	0.9	1.1	1.0	1.3	1.5	1.4	1.2	1.0	1.4
	P (mm)	−1.2	1.0	−5.2	−12.7	−10.6	−11.4	−9.8	−11.8	−6.8	−8.2	−8.7	17.8
	CO_2 (ppm)	550	550	550	550	550	550	550	550	550	550	550	550
12	T (°C)	1.6	1.5	2	1.7	1.9	2.1	2.5	2.7	2.4	2.8	1.6	2.3
	P (mm)	−26.0	−28.3	−12.9	−30.4	−15.5	−24.1	−17.2	−23.9	−30.7	−46.4	−37.9	54.7
	CO_2 (ppm)	550	550	550	550	550	550	550	550	550	550	550	550
13	T (°C)	1.8	1.6	1.6	1.7	2.2	2.5	2.6	3.0	2.7	2.5	2.2	2.2
	P (mm)	−3.5	−1.0	−5.2	−17.6	−31.8	−21.0	−14.5	−16.4	−19.1	−27.4	−8.7	−8.9
	CO_2 (ppm)	660	660	660	660	660	660	660	660	660	660	660	660
14	T (°C)	2.7	3.2	3.2	2.9	3.8	4.7	4.5	4.9	5.5	5.0	3.7	2.9
	P (mm)	36.6	17.2	−22.4	−26.5	−37.5	−30.2	−20.6	−28.5	−28.0	−56.1	−56.8	−29.6
	CO_2 (ppm)	660	660	660	660	660	660	660	660	660	660	660	660

Notes: Scenario 1, T + mean 2031–2060 C; Scenario 2, T + max 2031–2060 C; Scenario 3, T + mean 2069–2098 C; Scenario 4, T + max 2069–2098 C; Scenario 5, P − mean 2031–2060%; Scenario 6, P − max 2031–2060%; Scenario 7, P − mean 2069–2098%; Scenario 8, P − max 2069–2098%; Scenario 9, 1.5 × CO_2; Scenario 10, 2 × CO_2; Scenario 11, Scenario 1 + Scenario 5 + Scenario 9; Scenario 12, Scenario 2 + Scenario 6 + Scenario 10; Scenario 13, Scenario 3 + Scenario 7 + Scenario 9; and Scenario 14, Scenario 4 + Scenario 8 + Scenario 10.

Table 1. Climate change scenarios used in the simulations.

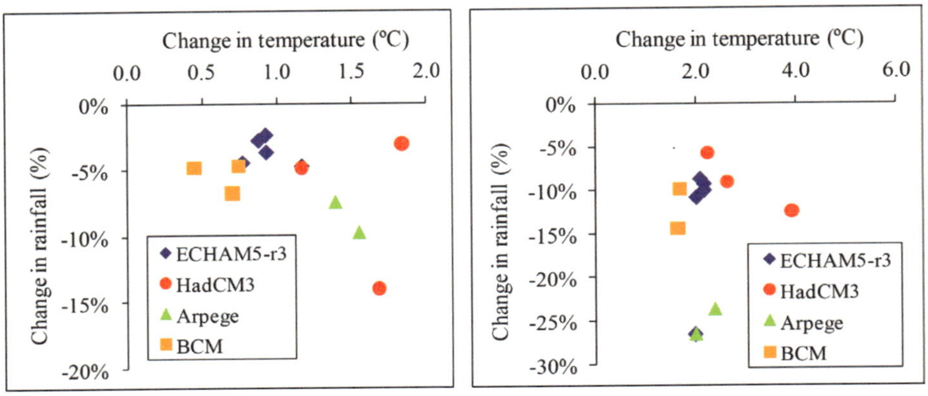

Figure 2. Variation range of forecast mean annual temperature and rainfall (ENSEMBLES project) for (a) period 2031–2060 and (b) period 2069–2098. Symbols identify different global models.

The different climatic scenarios used in this study are based on predicted future alterations from regional models in the ENSEMBLES project (socio-economic A1B scenario) for the closest meteorological station of the study area, for the periods 2031–2060 and 2069–2098. Due to variability of projections of future temperature and rainfall among the different models (**Figure 2**), the data from the models were combined to obtain the mean and maximum monthly rainfall and temperature for two periods: 2031–2060 (intermediate future) and 2069–2098 (distant future). Differences between projected and current values for the reference period (1981–2010) were used to develop the climate scenarios.

The stochastic weather generator has proven to be a useful tool for generating climate data series of high spatial and temporal resolutions to be used in climate change impact studies. In this study, the WXGEN weather generator included in the SWAT model was used to produce 30 years of synthetic daily weather data series for each climate change scenario, following the methodology used in [15, 17]. These weather data were used to run the SWAT model to simulate nitrate load under different climate change scenarios.

The nitrate load yields under the selected scenarios were compared with the 30-year simulation of the reference period. T-tests were conducted to determine the significance of the difference in nitrate load between the reference scenario and the climate change scenarios. All of the statistical tests were performed in PASW Statistics 18 for the Windows program package (SPSS Inc.) at a significance level of 0.05.

4. Results and discussion

4.1. Use of WXGEN weather

The SWAT model was previously calibrated and validated for streamflow, suspended sediment and nitrate yield in the study area [15, 27, 28]. Regarding the nitrate, it pointed out the importance of agricultural land (30% of catchment area) as the main contributor to N losses (77%), which reach the stream mainly in groundwater flow. The estimated yield (4.8 kg ha^{-1}) showed a close value to the measured values (5.1 kg ha^{-1}). Nash-Sutcliffe efficiency >0.50 and percent bias <10% were obtained during the calibration and validation period, indicating that the model was able to simulate the nitrate yield in the research area [28]. So, it was considered suitable for study of the impact of climate change on nitrate load in the catchment study.

The utility of the WXGEN weather generator embedded in the SWAT model was tested with the objective of assessing its use in simulations of climate scenarios. For this, the model was run using the climate generator for current conditions (reference period 1981–2010) to simulate nitrate load. Then, these results were compared with simulated nitrate load estimated using observed meteorological data. The statistical indicators (R^2 = 0.60 and NSE = 0.55), according to the criteria proposed by Refs. [29, 30], suggest a satisfactory model performance and indicate that the WXGEN weather generator can be used with a reasonable degree of confidence reasonable to analyse climate change scenarios in the study area.

4.2. Impacts of changes in temperature, rainfall and CO_2 concentrations in nitrate load

An increase in the nitrate load is predicted, both increasing in temperature and CO_2 concentrations, while a decrease was forecast for scenarios with reductions in rainfall (**Table 2**). The variation in the nitrate load in future scenarios is lower than that foreseen for streamflow and suspended sediment [15, 27], indicating thta it is less sensitive to changes in rainfall, temperature and CO_2 concentration than to discharge and sediment. The forecast pattern of nitrate load is similar to that streamflow, except in scenarios with changes in rainfall, as frequently

	NO_3 yield (kg ha^{-1} y^{-1})	**Percentage of change**	**NO_3 concentration (mg L^{-1})**
Reference period (1981–2010)	4.66		8.09
Scenario 1	4.89	5	9.71
Scenario 2	4.99	7	11.34
Scenario 3	5.01	8	12.28
Scenario 4	5.16	11	15.89
Scenario 5	4.67	0	9.18
Scenario 6	4.53	−3	10.70
Scenario 7	4.53	−3	10.71
Scenario 8	4.31	−7	15.06
Scenario 9	4.71	1	7.51
Scenario 10	4.75	2	7.18

Temperature (Scenarios 1–4), rainfall (Scenarios 5–8) and CO_2 concentration (Scenarios 9 and 10) based on scenarios defined in **Table 1**.

Table 2. Response of nitrate yield and concentration to changes in climate variables.

reported in the literature [12–14, 31], although this does not always happen. For example, Ficklin et al. [6] when analysing the sensitivity of nitrate load to increased CO_2 concentrations observed a decrease in nitrate yield linked to increased streamflow.

Since the entry of fertilisers into the simulations remained constant in relation to reference conditions, the forecast increase in nitrate load with increasing temperature is probably due to a greater contribution of N from agricultural areas, because of the decrease in plant biomass of grasslands and crops [15] with increasing temperature, as well as to the increase in organic nitrogen mineralisation. The N mineralisation in the soil depends on the nature and abundance of organic matter and temperature, humidity and pH and microbial activity. It is well known that it increases with the content of organic matter and temperature [32], which leads to an accumulation of inorganic nitrogen in the soil and an increased risk of leaching [33], provided that the water content does not limit the microbial activity [34]. In the study area, the annual rainfall is 1141 mm (1983/1984–2016/2017) so the water content of the soil should not be limiting for microbial activity, and, therefore, the increase of temperature could accelerate the transformation of nitrogen from organic to inorganic forms. Other authors, such as Ref. [31] in the Yorkshire river basin (the United Kingdom) and Ref. [14] in the Assiniboine basin (Canada), also attributed the increase in nitrate load to the accelerated mineralisation of biomass, although in all these cases, the nitrate followed the same trend as the streamflow.

The effects of climate variables on nitrate load were more noticeable on a seasonal level (**Figure 3**), highlighting the role of seasonal climate variations in affecting future nitrate. When changes in temperature were included, nitrate yield was forecast to rise in all seasons except summer, with the largest load increases in winter. These differences could be due to the increase of mineralisation in summer, with the consequent retention of nitrates in the soils because of the lack of water to transport them, while in the rainy seasons, the transport will be favoured, so it is more likely that load increases. Ref. [35] in laboratory experiences, carried out, therefore, with an artificial heating (heating, greenhouses), reported an increase in net mineralisation rates of 46%, while Ref. [36] when analysing the impact of climate change on quality of water in the Seine River (France) found an increase of between 8% and 26% in the net rate of mineralization.

When rainfall is modified, an increase in the nitrate load is expected in winter and a decline in the other seasons, especially in autumn, although the increase in winter does not compensate the losses in the other stations.

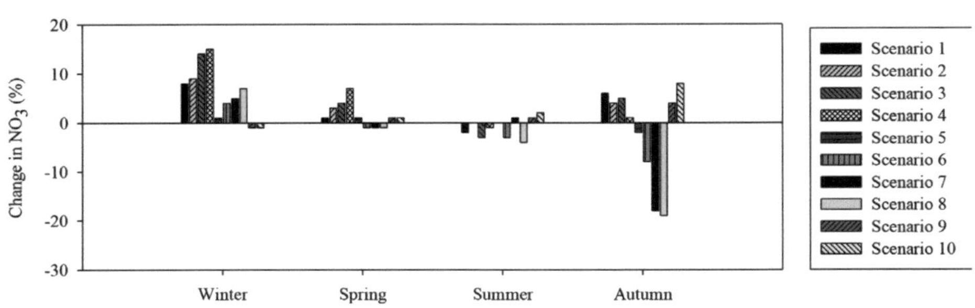

Figure 3. Seasonal response of nitrate load to changes in temperature (Scenarios 1–4), rainfall (Scenarios 5–8) and CO_2 concentrations (Scenarios 9 and 10) based on the scenarios defined in **Table 1**.

The results obtained at seasonal scale point out that the differences between seasons are attenuated for those scenarios that consider annual anomalies (e.g. Scenarios 3 and 4). This reveals that the scenarios that consider a certain exchange rate of temperature and/or rainfall, frequently used in the evaluation of impact of climate change on water resources and water quality, show the impacts on an annual scale but will hardly report processes that occur at smaller scales, because they do not take into account the distribution of temperatures and rainfall throughout the year, being necessary studies at seasonal scales.

4.3. Impacts of simultaneous changes in climate variables on nitrate load

Climate change is expected to increase the nitrate load in the Corbeira catchment (**Table 3**). This fact has been frequently attributted to greater water discharge [13, 14, 30]; however, Corbeira behaves differently in that there is an increase in nitrate load and a reduction in river flows, which shows that the streamflow will not be the determining factor of the nitrate load in this catchment in future scenarios. This contrast with the results of other studies in the Iberian Peninsula [12, 37], which reported reductions in N exports due to the decrease of streamflow.

The simulations performed with the average anomalies (Scenarios 11 and 13, **Table 1**) forecast an increase in the nitrate load in the order of 6% for the period 2031–2060 and 7% for the horizon 2069–2098, reflecting a great similarity for the entire twenty-first century, despite the notable differences expected in the streamflow, which will decrease by 16 and 35% at the mid and end of the twenty-first century [15]. An increase in the nitrate load during the spring and, especially, during the winter, which will be able to counteract the expected losses during the summer and autumn seasons (**Figure 4**), is observed. This behaviour could be related to an increased activity of the enzymes of the soil in the stations with greater water availability, as indicated by [38].

In general, nitrate losses depend on the hydrological balance, the quantities present in the soil (both from natural inputs and fertilisation) and the degree to which they are absorbed by vegetation [39]. It is known that rising temperatures and droughts exert a great influence on nutrient dynamics, since the warming increases mineralisation and drought prevents the absorption of nutrients from the plants and facilitates losses to the system when the rains arrive. The increase in the nitrate load with climate change, in this catchment, could be related to an increase in mineralisation and with the decreased nitrate absorption by

	NO_3 yield (kg ha^{-1} y^{-1})	Percentage of change	NO_3 concentration (mg L^{-1})
Reference period (1981–2010)	4.66		8.09
Scenario 11	4.94	6	10.16
Scenario 12	4.97	7	9.52
Scenario 13	4.97	7	20.12
Scenario 14	5.02	8	23.15

Table 3. Response of nitrate load to combined changes in temperature, rainfall and CO_2 concentrations based on the scenarios defined in **Table 1**.

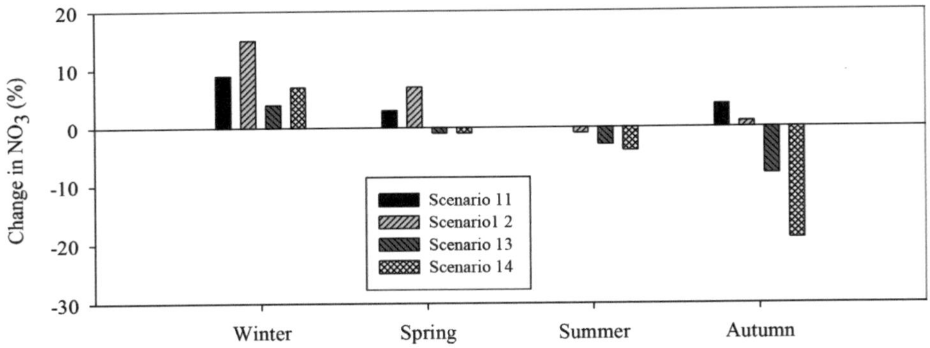

Figure 4. Seasonal response of nitrate load to combined changes in temperature, rainfall and CO_2 concentrations based on the scenarios defined in **Table 1**.

vegetation. A reduction in nitrate absorption is predicted for all land uses, especially significant in prairie areas (15% for the period 2031–2060 and 22% for 2069–2098). This points to less fertilizer being needed in these areas, which should be taken into consideration when planning management strategies in order to mitigate the impacts of potential climate change.

Despite the increase in nitrate concentration with climate change, the figures expected by the end of the twenty-first century would be well below the limits established by the current legislation for water for human consumption [40], so the supposed increase would be of concern for limitation for the human consumption, although it would result in degraded water quality.

The effects of land use have not been addressed in this study, since it was restricted to investigating the impacts of climate change on nitrate load because there is generally more uncertainty in climate projection than land use. So, more attention should be given to investigating the impacts due to climate change rather than land-use change. However, the effect of future land use on future nitrate load is controversial [10, 37, 39]. Some authors indicated that the effect of land cover is more visible than the climate change effect [37], while others found that stream nitrate concentrations were much more impacted by climate change than land-cover changes [10, 39]. This highlights the need to understand the combined effect of changes in land use and climate on catchment nitrogen discharge. This issue will be the aim of further research into modelling the water quality in the catchment study.

5. Summary and conclusions

This study was carried out to determine the effects of climate change on nitrate load in an agro-forested catchment located in NW Spain using the WXGEN weather generator included in the SWAT model. The results suggested that the WXGEN generator was able to adequately estimate long-term nitrate load.

Overall, it is verified that the nitrate load will increase in the future horizons in relation to current values (about 6 and 7% for the periods 2031–2060 and 2069–2098, respectively), possibly due to the decline in grassland biomass, as well as an increase in the rate of mineralisation linked to the increase in temperature. A higher rate of mineralisation of organic matter will result in the released nitrate being dragged by the water towards the fluvial course. Despite this, the concentrations of nitrates planned for the end of the twenty-first century would be well below the limits established by the current legislation in force for drinking water, so the supposed increase would not be a limitation for human consumption, although a deterioration in the quality of water in the study area is expected.

Acknowledgements

This paper is a contribution to the projects 10MDS103031 of the Xunta de Galicia and CGL2014-56907-R of the Programa Estatal de Investigación, Desarrollo e Innovación Orientada a los Retos de la Sociedad, which was funded by the Spanish Ministry of Economy and Competitiveness. M.L. Rodríguez-Blanco has been awarded a postdoctoral research contract (Juan de la Cierva Programme), which was funded by the Spanish Ministry of Economy and Competitiveness.

Conflict of interest

The authors declare no conflict of interest.

Author details

María Luz Rodríguez-Blanco*, María Mercedes Taboada-Castro, Ricardo Arias and María Teresa Taboada-Castro

*Address all correspondence to: mrodriguezbl@udc.es

Faculty of Sciences, Centre for Advanced Scientific Research (CICA), University of A Coruna, Coruña, Spain

References

[1] Wade A, Soulsby C, Langan SJ, Whitehead PG, Edwards AG, Butterfield D, et al. Modelling in stream nitrogen variability in the Dee catchment, NE Scotland. The Science of the Total Environment. 2001;**265**:229-252. DOI: 10.1016/S0048-9697(00)00661-6

[2] Kolenbrander GJ. Leaching of nitrogen in agriculture. In: Brogan JC, editor. Nitrogen Losses and Surface Runoff from Land Spreading of Manures. The Hague: Martinus Nijhoff; 1982. pp. 199-216

[3] Camargo JA, Alonso A, Salamanca A. Nitrate toxicity to aquatic animals: A review with new data for freshwater invertebrates. Chemosphere. 2005;**58**:1255-1267. DOI: 10.1016/j.chemosphere.2004.10.044

[4] Dupas R, Jomaa S, Musolff A, Borchardt D, Rode M. Disentangling the influence of hydroclimatic patterns and agricultural management on river nitrate dynamics from sub-hourly to decadal time scales. The Science of the Total Environment. 2016;**15**:791-800. DOI: 10.1016/j.scitotenv.2016.07.053

[5] Bindi M, Olesen JE. The responses of agriculture in Europe to climate change. Regional Environmental Change. 2010;**11**:151-158. DOI: 10.1007/s10113-010-0173-x

[6] Ficklin DL, Luo Y, Luedeling E, Gatzke SE, Zhang M. Sensitivity of agricultural runoff loads to rising levels of CO_2 and climate change in the San Joaquin Valley watershed of California. Environmental Pollution. 2010;**158**:223-234. DOI: 10.1016/j.envpol.2009.07.016

[7] Whitehead PG, Wilby RL, Battarbee RW, Kernan M, Wade AJ. A review of the potential impacts of climate change on surface water quality. Hydrological Sciences. 2012;**54**: 101-123. DOI: 10.1623/hysj.54.1.101

[8] Zessner M, Schönhart M, Parajka J, Trautvetter H, Mitter H, Kirchner M, et al. A novel integrated modelling framework to assess the impacts of climate and socio-economic drivers on land use and water quality. The Science of the Total Environment. 2017;**579**:1137-1151. DOI: 10.1016/j.scitotenv.2016.11.092

[9] EC. Directive 2000/60/EC of the European parliament and of the council of 23 October 2000 establishing a framework for community action in the field of water policy. Official Journal of the European Communities. 2000;**22**:72 L 327/1

[10] Gabriel M, Knightes C, Cooter E, Dennis R. Evaluating relative sensitivity of SWAT-simulated nitrogen discharge to projected climate and land cover changes for two watersheds in North Carolina, USA. Hydrological Processes. 2016;**30**:1403-1418. DOI: 10.1002/hyp.10707

[11] Steffen W, Richardson K, Rockström J, Cornell SE, Fetzer I, Bennett EM, et al. Planetary boundaries: Guiding human development on a changing planet. Science. 2015;**347**: 1259855-1-1259855-10. DOI: 10.1126/science.1259855

[12] Molina-Navarro E, trolle D, Martínez-Pérez S, Sastre-Merlín A, Jeppesen E. Hydrological and water quality impact assessment of a Mediterranean limno-reservoir under climate change and land use management scenarios. Journal of Hydrology. 2014:354-366. DOI: 10.1016/j.jhydrol.2013.11.053

[13] Martinkova M, Hesse C, Krysanova V, Vetter T, Hanel M. Potential impact of climate change on nitrate load from the Jizera catchment (Czech Republic). Physics and Chemistry of the Earth. 2011;**36**:673-683. DOI: 10.1016/j.pce.2011.08.013

[14] Shrestha RR, Dibike YB, Prowse T. Modeling climate change impacts on hydrology and nutrient loading in the upper Assiniboine catchment. Journal of the American Water Resources Association. 2011;**48**:74-89. DOI: 10.1111/j.1752-1688.2011.00592.x

[15] Arias R, Rodríguez-Blanco ML, Taboada-Castro MM, Nunes JP, Keizer JJ, Taboada-Castro MT. Water resources response to changes in temperature, rainfall and CO_2 concentration: A first approach in NW Spain. Water. 2014;**6**:3049-3067. DOI: 10.3390/w6103049

[16] Zabaleta A, Meaurio M, Ruiz E, Antigüedad I. Simulation climate change impact on runoff and sediment yield in a small watershed in the Basque Country, northern Spain. Journal of Environmental Quality. 2014;**43**:235-245. DOI: 10.2134/jeq2012.0209

[17] Meaurio M, Zabaleta A, Boithias L, Epelde AM, Sauvage S, Sánchez-Pérez JM, et al. Assessing the hydrological response from an ensemble of CMIP5 climate projections in the transition zone of the Atlantic region (Bay of Biscay). Journal of Hydrology. 2017;**548**:46-62. DOI: 10.1016/j.jhydrol.2017.02.029

[18] Álvarez V, Taboada JJ, Lorenzo MN. Cambio climático en Galicia en el siglo XXI: Tendencias y variabilidad en temperaturas y precipitaciones [Climate change in Galicia by the XXI century: Tendencies and variability in temperatures and precipitation]. Revista Avances Ciencias Terra (ACT). 2011;**2**:65-85

[19] IPCC (International Panel of Climate Change). Climate Change 2007: Synthesis report. In: Core Writing Team, Pachauri RK, Reisinger A, editors. Contribution of working groups I, II and III to the Fourth Assessment Report of the Intergovernmental Panel on Climate Change; Geneva. 2007. 103 p

[20] Instituto Tecnológico Geominero de España (IGME). Mapa Geológico de España, 1:50,000. Hoja 45. Madrid, Spain: Betanzos; Servicio de Publicaciones del Ministerio de Industria y Energía; 1981

[21] IUSS Working Group WRB. World Reference Base for Soil Resources 2014, Update 2015. International Soil Classification System for Naming Soils and Creating Legends for Soil Maps; World Soil Resources Reports No. 106. Rome, Italy: FAO; 2015

[22] Rodríguez-Blanco ML, Taboada-Castro MM, Palleiro L, Taboada-Castro MT. Temporal changes in suspended sediment transport in an Atlantic catchment, NW Spain. Geomorphology. 2010;**123**:181-188. DOI: 10.1016/j.geomorph.2010.07.015 479

[23] Rodríguez-Blanco ML, Taboada-Castro MM, Taboada-Castro MT. Rainfall runoff response and event based runoff coefficients in a humid area (northwest Spain). Hydrological Sciences Journal. 2012;**57**:445-459. DOI: 10.1080/02626667.2012.666351 476

[24] Rodríguez-Blanco ML, Taboada-Castro MM, Taboada-Castro MT. Linking the field to the stream: Soil erosion and sediment yield in a rural catchment NW Spain. Catena. 2013; **102**:74-81. DOI: 10.1016/j.catena.2010.09.003

[25] Arnold JG, Srinivasan R, Muttiah RS, Williams JR. Large area hydrologic modeling and assessment. Part I: Model development. Journal of the American Water Resources Association. 1998;**34**:73-89. DOI: 10.1111/j.1752-1688.1998.tb05961.x

[26] Neitsch SL, Arnold JG, Srinivasan R, Williams JR. Soil and Water Assessment Tool User's Manual. TX, USA: Texas Water Resources Institute: Colleague Station; 2002. 506 p

[27] Rodríguez-Blanco ML, Arias R, Taboada-Castro MM, Nunes JP, Keizer JJ, Taboada-Castro MT. Potential impact of climate change on suspended sediment yield in NW Spain: A case study on the Corbeira catchment. Water. 2016;**8**:444. DOI: 10.3390/w8100444 www.mdpi.com/journal/water

[28] Rodríguez-Blanco ML, Arias R, Taboada-Castro MM, Nunes JP, Keizer JJ, Taboada-Castro MT. Modelling the contribution of land use to nitrate yield from a rural catchment. Landscape Ecology. 2016;**I**:27-28. DOI: 10.5772/63718

[29] Motovilov Y, Gottschalk GL, Engeland K, Rodhe A. Validation of distributed hydrological model against spatial observations. Agricultural and Forest Meteorology. 1999;**98**: 257-277. DOI: 10.1016/S0168-1923(99)00102-1

[30] Moriasi DN, Arnold JG, van Liew MW, Bingner RL, Harmel RD, Veith TL. Model evaluation guidelines for systematic quantification of accuracy in watershed simulations. Translations ASABE. 2007;**50**:885-900. DOI: 10.13031/2013.23153

[31] Bouraoui F, Galbiati L, Bidoglio G. Climate change impacts on nutrient loads in the Yorkshire Ouse catchment (UK). Hydrology and Earth System Sciences. 2002;**6**:197-209. DOI: 10.5194/hess-6-197-2002

[32] Leirós MC, Trasar-Cepeda C, Seoane S, Gil-Sotres F. Dependence of mineralization of soil organic matter on temperature and moisture. Soil Biology and Biochemistry. 1999;**31**: 327-335. DOI: 10.1016/S0038-0717(98)00129-1

[33] Olesen JE, Bindi M. Consequences of climate change for European agricultural productivity, land use and policy. European Journal of Agronomy. 2002;**16**:239-262. DOI: 10.1016/S1161-0301(02)00004-7

[34] Zak DR, Holmes WE, MacDonald NW, Pregitzer KS. Soil temperature, matric potential, and the kinetics of microbial respiration and nitrogen mineralization. Soil Science Society of America Journal. 1999;**63**:575-584. DOI: 10.2136/sssaj1999.03615995006300030021x

[35] Rustad LE, Campbell JL, Marion GM, Norby RJ, Mitchell MJ, Hartley AE. A meta-analysis of the response of soil respiration, net nitrogen mineralization, and aboveground plant growth to experimental ecosystem warming. Oecologia. 2001;**126**:543-620. DOI: 10.1007/s004420000544

[36] Ducharne A, Baubion C, Beaudoin N, Benoit M, Billen G, Brisson N. Long term prospective of the Seine River system: Confronting climatic and direct anthropogenic changes. The Science of the Total Environment. 2007;**375**:292-311. DOI: 10.1016/j.scitotenv.2006.12.011

[37] Carvalho-Santos C, Nunes JP, Monteiro AT, Heins L, Honrado JP. Assessing the effects of land cover and future climate conditions on the provision of hydrological services in a medium sized watershed of Portugal. Hydrological Processes. 2016;**30**:720-738. DOI: 10.1002/hyp.10621

[38] Sardans J, Peñuelas J, Estiarte M. Changes in soil enzymes related to C and N cycle and in soil C and N content under prolonged warming and drought in a Mediterranean shrubland. Applied Soil Ecology. 2008;**39**:223-235

[39] Ferrier RC, Whitehead PG, Sefton C, Edwards AC, Puhg K. Modeling impacts of land use change and climate change on nitrate nitrogen in the River Don, North East Scotland. Water Research. 1995;**29**:1950-1956

[40] European Council. Directive 75/440/CEE of the 16 June 1975 for provision of water to provide potable water. Official Journal of the European Union L. 1975;**194**:26

Chapter 3

Tools and Methods for Supporting Regional Decision-Making in Relation to Climate Risks

Jyri Hanski, Jaana Keränen and Riitta Molarius

Additional information is available at the end of the chapter

http://dx.doi.org/10.5772/intechopen.80322

Abstract

Climate change has had a major impact on the Nordic region. For example, the mean temperature rise is expected to be 4–6°C by 2080. In Finland, the regional authorities are responsible for climate change adaptation. Some of the most vulnerable sectors include energy, tourism, transport and water supply. Currently, it appears that the authorities are not familiar with the tools for assessing climate risks and lack knowledge about the impact of climate change. In this paper, we provide a review of risk assessment methods and decision-making tools, focusing on adapting to climate change in a Finnish context. Our research method comprises a systematic qualitative literature review dealing with relevant journals, dissertations and deliverables of relevant EU projects since 2005.

Keywords: climate change, climate change adaptation, decision-making, land-use planning, literature review, local authority, Nordic, risk assessment, risk assessment tools

1. Introduction

Finland is located north of the 60th parallel, making it one of the northernmost countries in the world. It has been predicted that the increase in temperature in the northern hemisphere as a result of climate change will be faster and higher than average, and winter temperatures in particular will rise with increasing precipitation. This means milder winters with less sun, less snow but more rain [1]. The mean temperature rise is expected to be 4–6°C by 2080. Even though Finland will probably not be affected by major floods or long-term heat waves, there are still many climate change impacts which need to be adapted. The change from frozen land to unfrozen land during the winter will be particularly challenging for many sectors. For example, the increased presence of unfrozen ground requires plant breeding to develop grain varieties that can withstand shorter winter precipitation and longer, perhaps drier summers [2]. In the forestry sector, winter storms and unfrozen land expose spruce forests to storm damage as the spruces' roots are torn from the ground [3]. Fallen trees may also sever power lines or railway catenary causing disruptions in the electricity supply [4, 5]. A rainy winter with no deep-rooted vegetation exposes roads and railways to erosion faster than ever before [6]. Tourism in Lapland may also suffer from an earlier spring season and warmer winters with less snow [7]. Finnish water utilities distribute 60% of their groundwater volume, which is mainly potable without any need for purification [8]. However, increasing precipitation may dilute the quality of groundwater and increase the cost of purification.

In Finland, primary responsibility for improving adaptation to climate change lies with the Ministry of Agriculture and Forestry, which published the National Adaptation Plan for Climate Change 2022 in 2014. This paper details the goals and objectives of adaptation activities, as well as the main measures and players. Usually, in the real world, the authorities are the main stakeholders for responding to adaptation for climate change. Thus, the national research institutes have been tasked with studying the effects of climate change and developing new means of adaptation. For example, the Finnish Environmental Centre is studying environmental tolerance, the Natural Resources Institute is studying forest and plantation areas, the Geological Survey is studying groundwater and the Finnish Meteorological Institute's (FMI) role is to predict the future climate. All these institutes provide up-to-date information to municipal, regional and state authorities in order to adapt to climate change. Indeed, the FMI has created websites that provide information on a regional level for decision-makers and the general public about the effects of climate change in different parts of Finland (see [1]).

Municipalities have a main role in adapting to climate change as they are able—through land-use planning and building regulations—to decide on where to build, how to build, what kind of response needs to be arranged, what kind of transport network to use, etc. [9]. However, adaptation to climate change also requires co-operation between various sectors and levels of administration. In Finland, it has been stated that the adaptation policy should be mainstreamed and integrated to fully cover public administration, and co-operation with the private sector and the third sector players should also be developed [10]. Land use in particular needs to be reviewed as a cross-sectoral issue.

Adapting to climate change involves multiple strategies, for example, reducing the sensitivity of the system by increasing the safety margin of new investments or using reversible options by trying to keep cost as low as possible (see [11, 12]). Whatever strategy is used, it is important to select case-specific risk assessment approaches to ensure adequate risk management measures [13].

Several methods have been developed over the last decade for supporting decision-making in relation to climate change. In this paper, we assess the latest methods that could be used to support regional or municipal decision-making, especially in Finland. This paper studies these methods and classifies them to help decision-makers select adequate methods for their purposes.

2. Methodology

This paper utilises a comparative analysis of climate risk assessment methods and frameworks based on a systematic literature review. A systematic review provides an audit trail of the reviewers' decisions, procedures and conclusions [14, 15]. It adopts a replicable, scientific and transparent process that aims to minimise bias through an exhaustive literature search. In this paper, an iterative process that has been modified from methods presented in [14, 16] has been utilised: (1) data collection and (2) descriptive analysis and data evaluation.

In this paper, climate risks are considered as being risks that result from climate change and that affect natural and human systems and regions. The risk assessment tool is a tool for assessing risks, that is, to determine a quantitative or qualitative estimate of risk related to a well-defined situation and a recognised threat.

2.1. Data collection

In this phase, the data to be collected were defined and delimited. The systematic literature review was conducted using the eKnowledge database, which enables access to a large number of scientific databases such as Scopus, Web of Science, ScienceDirect, and open access databases. Keywords for the search comprised a combination of "blizzard," "climate," "climate change," "climate risk," "climate risk assessment," "cold spell," "decision," "decision support," "extreme event," "extreme weather," "heat wave," "heavy precipitation," "heavy rain," "Nordic," "risk assessment," "risk assessment method," "storm," "tourism" and "wind gust." The literature review was complemented by methods and frameworks with which the authors are already familiar (**Figure 1**).

The preliminary screening was conducted based on the titles and abstracts of the papers. Journal-specific screening was also simultaneously performed including journals such as

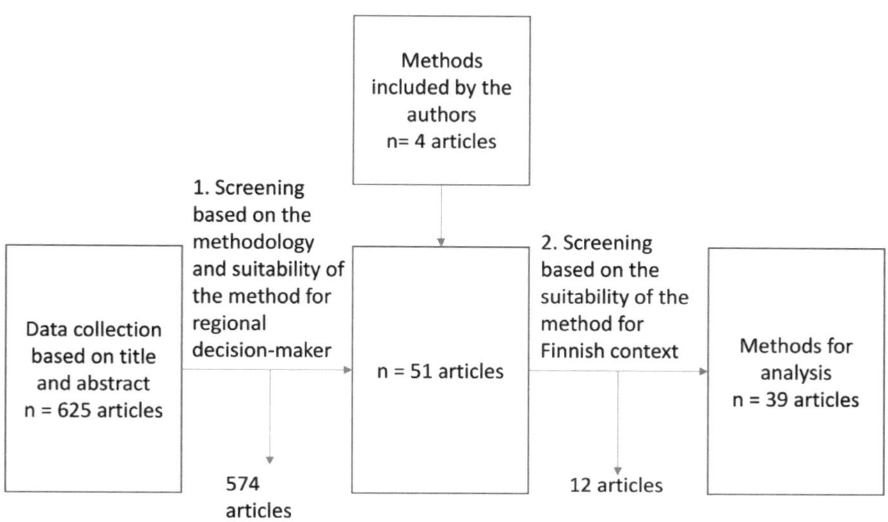

Figure 1. The data collection process.

Climate, Climate Services, Coastal Engineering, Geographia Napocensis, Natural Hazards, Ocean and Coastal Management, Science of the Total Environment, Transportation Research Procedia and Water Resources Management. Papers published between 2005 and 2018 were included in the review and only included the most recent papers. We focused on papers that described climate risks that were considered important in a Finnish context. Additionally, we concentrated on the sectors and industries that are important to the Finnish economy and most affected by climate change. Peer-reviewed articles, conference papers and book chapters were included to provide rich material for analysis. At this point, we focused on methods that could be easily used by regional decision-makers. Thus, methods that require expertise in order to use complex models were not included. Additionally, portfolio theory, real options and methods utilising future study methods were excluded (pure scenario methods, some methods with scenario components were included). After the first screening, 51 different papers were selected for the next stage.

In the second stage, the criteria that took into account the Finnish context were used. These included the next issues:

- The development of flood risk maps has not been included in the research as these kinds of maps are already available in Finland from the authorities. However, if there were ways of upgrading these kinds of maps, they were included in the further analysis.
- Methods that study the rise in sea level have not been taken into account because, in Finland, it is believed that the rise in ground level since the last Ice Age is still higher than the rise in sea level in most part of the country [1]. However, the methods that address storm floods and the methods that integrate the rise in sea level into more extensive methods have been taken into account.
- The methods planned for areas with scarce data have not been taken into account because the whole of Finland has been covered by an effective weather monitoring network since 1880.
- Articles that only deal with aspects of resilience have been excluded.
- Methods that can be performed with minor assistance from consultancy services have been taken into consideration. For example, climate change projections, flooding predictions, or groundwater level variations are examples of data in which an expert may be required to facilitate the interpretation. These, in turn, are often the starting point for the many existing methods such as risk maps.

After the second screening, 39 papers were selected for further analysis. The papers included in the literature review are presented in Notes.

2.2. Descriptive analysis and data evaluation

The formal aspects of the data were assessed and categories were selected and applied to the collected data during the descriptive analysis and category selection. Eight categories of climate risk assessment and decision support methods were identified. Additionally, information on the basic characteristics of the method, such as location, phenomenon or risk, application and time frame for analysis, was classified. These categories were used in the subsequent evaluation.

During the data evaluation phase, the material was thematically analysed according to the selected categories. The validity and reliability of the results were increased by using an iterative process. When analysing the data, the authors looked for emerging classifications and patterns. The classifications were created based on the classifications used in the literature and findings from the data.

3. Decision-making in the context of climate risks

There is a myriad of methods available for supporting climate risk assessment and decision-making. The nature of methods can vary greatly regarding data requirements, time frame and purpose [17]. UNFCCC [18] has identified the following approaches and methods: cost-benefit analysis (CBA), cost-effectiveness analysis (CEA), multi-criteria decision analysis (MCDA), portfolio theory and real options, pathway analysis, adaptive capacity assessment, risk management methods, scenario-based approaches, technological assessments, normative policy assessments, identifying learning in individuals/organisations, participatory methods and social learning.

These methods are mostly complementary in nature and can be used to support a variety of different climate change-related decision-making situations. In this paper, we introduce a group of climate risk assessment and decision support methods that we consider suitable for a regional decision-maker.

3.1. Cost-benefit analysis

Cost-benefit analysis (CBA) can be defined as a comparison of the marginal costs of policies with the marginal benefits associated with the climate change effects that are prevented in order to identify the most economically efficient policy response [19]. It provides monetary valuations for every kind of impact involved and is particularly suited to supporting decisions related to the feasibility of investment projects, in which the future financial effects can be identified and predicted [20]. It is considered a more objective method compared to its main competitors, MCDA and CEA [21]. However, there are issues in using CBA for climate risk assessment. Multiple externalities are difficult to value and do not figure in the evaluation of costs and benefits [20], and the inclusion of complex features such as future time, doubt, irreversibility and indirect benefits is difficult [22, 23].

3.2. Cost-effectiveness analysis

Scrieciu et al. [20] defined cost-effectiveness analysis (CEA) as an identification of least-cost options to meet a certain target or policy objective. The rationale behind CEA is that there is a single indicator of effectiveness. Cost curves are a classic application area of CEA. CEA has been criticised for the difficulties it has in identifying consistent metrics for adaptation and the local- and sector-specific nature of climate impacts [23].

3.3. Multi-criteria decision analysis

In a complex decision-making situation involving multiple stakeholders (i.e. climate risk-related decisions), a decision-maker may have several conflicting objectives. Multi-criteria decision analysis (MCDA) permits the consideration of quantitative and qualitative data together using multiple decision criteria [18]. With MCDA, the benefits and costs are measured on a value scale that reflects the desirability of the options from the perspective of the decision-maker [24]. The decision criteria should reflect which features decision-makers find important in decision-making [25]. Weights are given to each criterion, and the weighted sum of the different criteria is taken in order to gain an overall score for option, which, in turn, can be used to rank options [23]. The use of MCDA is appropriate when it is difficult to assign monetary value to the decision criteria. However, some of the same critique applies to MCDA as CBA and CEA [23].

3.4. Robust decision-making

In robust decision-making (RDM), the goal is to identify the full range of plausible future states and make decisions that are robust across a wide range of such future states as possible [20]. It starts with selecting decision options and then estimates utilities of options to identify the potential vulnerabilities of strategies [23]. RDM provides an analytical decision support framework for situations characterised by high uncertainty. Four key elements of RDM include: (1) assembling a high number of scenarios, (2) seeking robust strategies that perform sufficiently well across a broad range of futures, (3) employing adaptive strategies to achieve robustness and (4) designing an analysis for interactive exploration of the plausible futures [23]. Issues related to RDM include the complexity of the method and the need for advanced statistical and mathematical methods [20].

3.5. Participatory methods

Assessing climate risks often requires an approach that incorporates the perspectives of stakeholders in the problem and solution definition. Participatory methods cover a variety of approaches that support the inclusion of experts and users in the decision-making and assessment process (see e.g. [23]). Participatory methods are often utilised in methods such as MCDA to provide weights and valuations for criteria that are difficult to otherwise quantify. As standalone methods, they are utilised, for example, in understanding complexity, participatory analysis and stakeholder engagement and mapping. It is argued that participatory methods based on the joint work of scientists, experts and stakeholders lead to better assessments because they combine the latest expert information with first-hand policy experience in the affected society [26].

3.6. Risk analysis methods

The risk analysis process is presented in the risk management standard [27]. According to this standard, risk analysis forms part of a broader risk assessment process focusing on the nature of risk and its characteristics. Risk analysis involves a detailed consideration of uncertainties, risk sources, consequences, likelihoods, events, scenarios, controls and their effectiveness. In practice, detailed and diverse information is not always available. Even so, one main principle of risk analysis is to use the best available information, which is supplemented during the process.

Highly uncertain events can be difficult to quantify. This can be a disadvantage if, for example, low probability but high cost events are analysed. In such cases, a combination of methods may result in a better understanding.

3.7. Event tree analysis

The aim of Event tree analysis is to identify the undesirable consequences of an initial harmful event. ETA focuses on identifying the failure combinations that could lead to undesirable outcomes [28]. Because of the tree structure, it is possible to assign probabilities to these outcomes. This method has been used to model weather-induced event chains [28–30].

3.8. Risk index methods

Risk index methods stipulate methods that provide a numeric value (index) for the identified risks. The calculations are based on several factors that impact the risk, which are often categorised in order to obtain comparable values. The equations for calculation may include long impact chains, as in the case of the EWRI index [13]: $R = f(H,V)$; $H = f(P)$; $V = f((ExS)/CC)$. In this chain, R = risk, H = hazard, V = vulnerability, P = probability, E = exposure, S = susceptibility and CC = coping capacity.

4. Results

The analysed methods suitable for a Finnish context were categorised into different classes. One classification was based on the climate risk to which the method was applied (see **Figure** 2). The most common climate phenomenon was flood. Other climate risks such as storms or extreme weather events were also included in the flood risk examination. Two methods were applied to winter storm hazards. Heat wave or drought risks were examined using two methods. Eight methods were applied to multivariable risks such as landslide, drought, flood, sea level rise and erosion. In connection with seven methods, the applicable climate risks were not specified.

Another classification was the application area in which the method in question was field tested (see **Figure 3**). The most common application area comprised infrastructure in which planning applications in general and the identification of vulnerable and valuable assets not specified in more detail was also classified. Five methods were tested in the energy sector. In addition, water supply or water management comprised the application areas for eight methods and transportation comprised the application area for four methods. Only two methods focused on the tourism sector. Five methods were described without mentioning a specific application area.

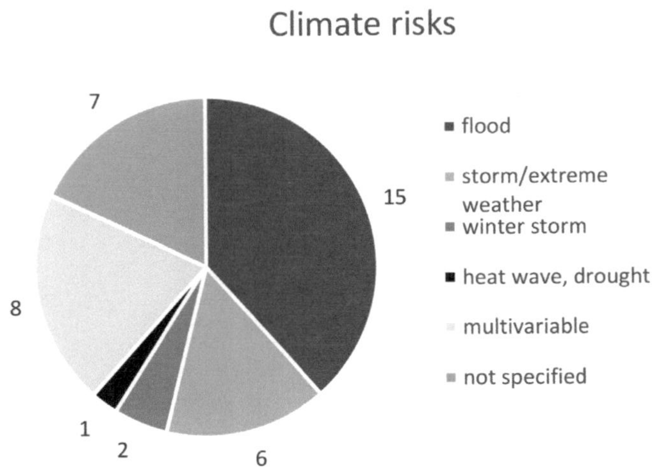

Figure 2. Classification of methods according to the climate risks concerned.

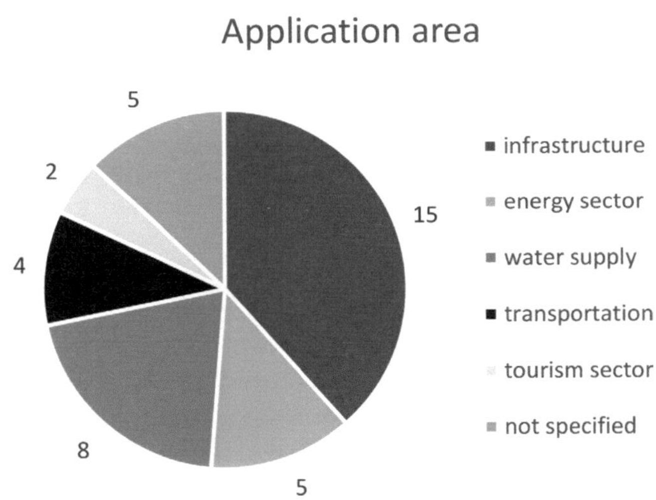

Figure 3. Classification of methods according to application areas.

Most of the methods applied to medium- or long-term planning assisting strategic (5–15 years), infrastructure (15–70 years) or land-use planning (over 50 years). One third of the analysed methods were suitable for short-term operational planning (0–5 years), while only four methods were regarded as being suitable for analysing risks 70 years into the future.

Almost all methods that were hypothesised beforehand were evaluated suitable for regional decision-makers' use. Only methods in which cost-effectiveness analysis (CEA) was applied in a suitable way compared to Finnish content criteria were not found. Various visual risk assessment tools, for example, risk or vulnerability maps, were identified. Visual tools were not previously described as a method for carrying out a climate risk assessment. Thus, to highlight their relevance and abundance, these tools are presented as a separate group.

4.1. Cost-benefit analysis (CBA)

Five articles contained CBA approaches. Three articles discussed flood risk assessment [31–33] and two discussed the adaptation of the electricity sector to climate change and extreme weather events [34, 35]. The methods supported infrastructure (public infrastructure, electricity and the transport sector) adaptation. The methods supported a variety of decision-making situations and time frames from strategic planning (5–15 years) to infrastructure and land use (15–70+ years).

4.2. Multi-criteria decision analysis (MCDA)

MCDA was used in nine articles. Eight of the articles dealt with flooding and coastal risks (e.g. flooding, storms and erosion) and their effect on infrastructure and land use [24, 36–42]. Two of the articles also considered other events such as heat wave, drought, wildfire and windstorm. One article focused on energy sector adaptation [43]. All of the methods supported strategic, infrastructure or land use decision-making. However, three of the articles were also intended for operational decision-making support (0–5 years) [38, 39, 43].

4.3. Robust decision-making (RDM)

Three articles utilised the RDM approach. All of the articles considered the adaptation of the infrastructure sector, specifically water sector adaptation [44–46]. Factors considered in the papers included, for example, climate conditions, water demand, systems operation and cost-related uncertainties.

4.4. Participatory methods

Five articles included a participatory method approach. Three of the articles focused on specific climate impacts such as flood, storm and landslide [47–49], while two of the articles were more general approaches to climate risk assessment [50, 51]. The time frame of the presented methods varied from covering tools to obtaining information for operational planning to supporting long-term infrastructure planning. Diverse methods were introduced in the papers to collect bottom-up information, for example, a gamified assessment method, web-based participatory methods and more traditional focus group meeting methods.

4.5. Vulnerability or risk assessment

Six articles were established based on vulnerability or risk assessment methods. The vulnerability assessment method focused on the tourism sector and studied the vulnerability of cross-country skiing to climate change impacts [52]. Two risk assessment methods examined storm risks in coastal areas [53, 54]: one studied risks to groundwater and related ecosystems [24] and one studied risk assessment methods for the road infrastructure and transport [55]. One method analysed future risks to hydropower plants based on climate scenarios [18]. The methods also utilised visual tools such as exposure [54], vulnerability [52] or hazard maps [53].

4.6. Event tree analysis

ETA appears to be a straightforward method for modelling the direct consequences of the impact chains of weather events. It is recommended that the method is used in two stages: firstly, the risk analysis team specifies the generic event tree model including its main branches, and secondly, sector-specific experts are asked to complete it by providing probabilities for each alternative branch [30]. ETA was utilised for flood risk management [30] and electricity infrastructure adaptation to snow storm effects [56].

4.7. Risk index methods

The group of risk index methods includes both index calculations and key performance indicators (KPI) of harmful weather events. For example, EWRI (extreme weather risk index) is based on the probability of a weather event and the vulnerability of transport routes [57]. Also, the method that deals with KPIs assesses the risk of climate change and presents the results in a visual format [43]. These methods are based on mathematical risk functions.

4.8. Maps or other visual tools

Seven articles represented maps or other visual tools [13, 58–63]. Most of the methods in this category related to expected future changes in water resources, such as rising seawater, groundwater level variation, flooding or other extreme water flow events in a map format. Some of the maps primarily projected hydrology changes in hydrological cycle-like flood maps [61], and some were combinations integrating both water supply and demand scenarios [59]. A number of the maps were made to cover a wide area such as national-level representations, while parts of the maps were considerably more high resolution with regard to a particular river basin or district [63]. There was some variation in time frames but most of the methods focused on strategic planning or planning of infrastructure or land use. It is also noticeable that methods classified in other decision-making support groups sometimes included visual tools. For example, flood maps were utilised as part of the process.

5. Discussion

Many different methods were identified that were suitable for regional decision-making related to climate change adaptation and climate risk management. Apart from cost-effectiveness analysis (CEA), all other methods, which were previously hypothesised as being applicable to decision-making, came up during the literature review. Articles concerning CEA were also identified, but the presented methods did not fit the inclusion criteria. With regard to risk assessment methods, it appears that the main focus of recent research has been on studying the environmental impacts of climate change. These studies provide impact models that primarily concern water levels, drought, precipitation, wind gusts, etc.

Veijalainen [64] has described the chain from a global climate scenario to its environmental impacts as demonstrating that information from the climate scenario must be converted into

http://dx.doi.org/10.5772/intechopen.80322

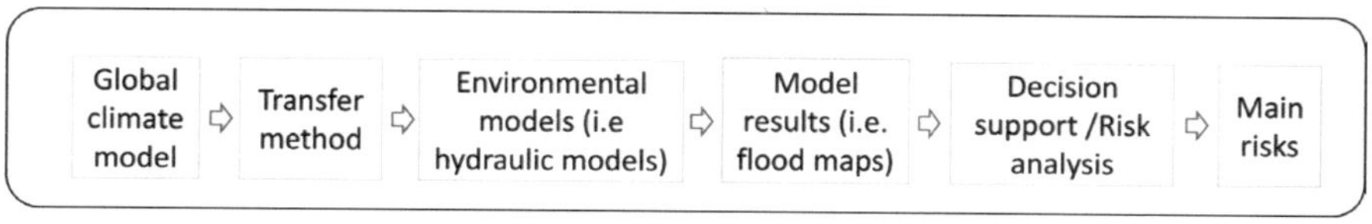

Figure 4. Converting global climate model information into regional or municipal decision-making.

information on hydraulic models. These models can be used for evaluating, for example, the return period of floods. Only then will municipal or regional authorities be able to carry out a risk assessment to determine what kind of adaptation methods they can select (**Figure 4**).

It appears that most of the studied articles use the term "risk assessment" to describe the analysis and methods of converting information from a global climate scenario for environmental models, such as flood models, evaporation models, etc. These models are suitable for a scenario analysis and are used for adapting to climate change from a long-term perspective in which the planning period is more than 30–50 years. The results of these methods require more specific risk assessment to support municipal or regional decision-making in a shorter time frame.

In addition to expertise on climate change, there is a need for further information on local or regional vulnerability. For example, the vulnerability of infrastructure, sensitive assets or socio-technical systems generally has to be taken into account in some way when analysing regional or local level climate risks and potential impacts.

The categorisation of methods was sometimes challenging. Some of the methods could have been categorised into several different classes. For example, the differences between risk analysis and multi-criteria decision analysis methods were not always obvious. In addition, the applicability of the method for a specific temporal extent (operational, strategic or land-use planning) was not obvious every time if it was not specified.

In order to find out the suitability of the methods in a Finnish context, a more detailed analysis should be performed. For example, the applied climate and hydrology models may be suitable for Finnish context, but only under case-specific circumstances. Also, the Baltic Sea is not included in most global climate models, even though it significantly influences the Finnish climate. Applications should be taken into account when adapting methods in new types of geographical or case areas. In Europe, climate risks, application areas and infrastructure are quite similar or, at least, consistent characteristics exist, and knowledge of climate and hydrology models, historical data, hazard events, infrastructure and current socio-technical systems also exist. On a general level, methods are applicable to various geographical areas and decision contexts.

6. Conclusions

The importance of climate change adaptation has been identified on a regional level in Finland. This chapter focused on methods of climate risk assessment that are suitable for regional decision-makers. Using a systematic literature review, 39 methods were identified that could support regional decision-makers. A wide range of methods were identified including multi-criteria methods, methods of analysing costs, the benefits of different options, risk assessment methods and the methods that utilise visual tools. The methods highlighted climate risks linked to hydrological cycles such as storm-induced risks and flood risks. However, the majority of the identified methods require consultancy assistance. Most of the methods include, for example, climate change projections or hydrology models that are quite complex and require specific knowhow in order to be applied in a case-specific manner.

Conflict of interest

The authors certify that they have no affiliations with or involvement in any organisation or entity with any financial interest (e.g. honoraria; educational grants; participation in speakers' bureaux; membership, employment, consultancies, stock ownership, or other equity interest; and expert testimony or patent-licensing arrangements) or non-financial interest (such as personal or professional relationships, affiliations, knowledge or beliefs) in the subject matter or materials discussed in this manuscript.

Notes

Papers included in the systematic literature review: Refs. [13, 24, 30–67].

Author details

Jyri Hanski*, Jaana Keränen and Riitta Molarius

*Address all correspondence to: jyri.hanski@vtt.fi

VTT Technical Research Centre of Finland Ltd., Tampere, Finland

References

[1] FMI. Climate-Guide [Internet]. 2018. Available from: https://ilmasto-opas.fi/en/ilmastonmuutos/suomen-muuttuva-ilmasto/-/artikkeli/338246aa-d354-4607-b087-cd9e0d4a3d04/maankohoaminen-hillitsee-merenpinnan-nousua-suomen-rannikolla.html [Accessed: 15-05-2018]

[2] Peltonen-Sainio P, Sorvali J, Müller M, Huitu O, Neuvonen S, Nummelin T, et al. Adaptation 2017 in Finland [in Finnish: Sopeutumisen tila 2017] the Natural Resources Institute Report 18/2017. 2017. Available from: http://jukuri.luke.fi/bitstream/handle/10024/538722/luke-luobio_18_2017.pdf?sequence=1&isAllowed=y [Accessed: 15-05-2018]

[3] Kirkinen J, Martikainen A, Holttinen H, Savolainen I, Auvinen O, Syri S. Impacts on the Energy Sector and Adaptation of the Electricity Network Business Under a Changing Climate in Finland. FINADAPT Working Paper 10. 2005. Available from: http://hdl.handle.net/10138/41052 [Accessed: 18-06-2018]

[4] Forssén K, Mäki K, Räikkönen M, Molarius R. Resilience of electricity distribution networks against extreme weather conditions. ASCE-ASME Journal of Risk and Uncertainty in Engineering Systems, Part B: Mechanical Engineering RISK-16-1133. 2017;**3**(2). DOI: 10.1115/1.4035843

[5] Molarius R, Räikkönen M, Forssén K, Mäki K. Enhancing the resilience of electricity networks by multi-stakeholder risk assessment: The case study of adverse winter weather in Finland. Journal of Extreme Events. 2016;**3**(4):1650016. DOI: 10.1142/S2345737616500160

[6] Peltonen L, Haanpää S, Lehtonen S. The Challenge of Climate Change Adaptation in Urban Planning. FINADAPT Working Paper 13. Finnish Environment Institute. 2005. Available from: http://hdl.handle.net/10138/41059 [Accessed: 18-06-2018]

[7] Landauer M, Sievänen T, Neuvonen M. Indicators of climate change vulnerability for winter recreation activities: A case of cross-country skiing in Finland. Leisure/Loisirs. 2015;**39**(3-4):403-440. DOI: 10.1080/14927713.2015.1122283

[8] GTK [Internet]. 2018. Available from: http://www.gtk.fi/geologia/luonnonvarat/pohjavesi/ [Accessed: 18-06-2018]

[9] Kuntaliitto. Municipalities in Changing Climate [in Finnish; Kunnat ilmastonmuutoksessa] Association of Finnish Local and Regional Authorities. 2018. Available from: https://www.kuntaliitto.fi/sites/default/files/media/file/ilmasto_ebook.pdf [Accessed: 18-06-2018]

[10] Mickwitz P, Kivimaa P, Hildén M, Estlander A, Melanen M. Ilmastopolitiikan valtavirtaistaminen ja politiikkakoherenssi. Selvitys Vanhasen II hallituksen tulevaisuusselontekoa varten. Valtioneuvoston kanslia, Helsinki. Valtioneuvoston kanslian julkaisusarja 6/2008. 2008;**74**. Available from: http://docplayer.fi/56252278-Ilmastopolitiikan-valtavirtaistaminen-ja-politiikkakoherenssi-valtioneuvoston-kanslian-julkaisusarja.html [Accessed: 18-06-2018]

[11] Miller S, Muir-Wood R, Boissonade A. An exploration of trends in normalized weather-related catastrophe losses. In: Diaz HF, Murnane RJ, editors. Climate Extremes and Society. Cambridge, UK: Cambridge University Press; 2008. pp. 225-241. DOI: 10.1017/CBO9780511535840.014

[12] Hallegatte S. Strategies to adapt to an uncertain climate change. Global Environmental Change. 2009;**19**:210-247. DOI: 10.1016/j.gloenvcha.2008.12.003

[13] Molarius R, Keränen J, Poussa L. Combining climate scenarios and risk management approach—A Finnish case study. Climate. 2015;**3**(4):1018-1034. DOI: 10.3390/cli3041018

[14] Tranfield D, Denyer D, Smart P. Towards a methodology for developing evidence-informed management knowledge by means of systematic review. British Journal of Management. 2003;**14**(3):207-222. DOI: 10.1111/1467-8551.00375

[15] Cook DJ, Greengold NL, Ellrodt AG, Weingarten SR. The relation between systematic reviews and practice guidelines. Annals of Internal Medicine. 1997;**127**(3):210-216. DOI: 10.7326/0003-4819-127-3-199708010-00006

[16] Seuring S, Müller M. From a literature review to a conceptual framework for sustainable supply chain management. Journal of Cleaner Production. 2008;**16**:1699-1710. DOI: 10.1016/j.jclepro.2008.04.020

[17] Raseman WJ, Kasprzyk JR, Rosario-Ortiz FL, Stewart JR, Livneh B. Emerging investigators series: A critical review of decision support systems for water treatment: Making the case for incorporating climate change and climate extremes. Environmental Science: Water Research & Technology. 2017;**3**:18-36. DOI: 10.1039/c6ew00121a

[18] UNFCCC. Potential Costs and Benefits of Adaptation Options: A Review of Existing Literature [Internet]. United Nations. 2010. Available from: https://unfccc.int/resource/docs/2009/tp/02.pdf [Accessed: 20-06-2018]

[19] Dessler AE, Parson EA. The Science and Politics of Global Climate Change: A Guide to the Debate. Cambridge: Cambridge University Press; 2006. ISBN-10: 0-521-53941-2

[20] Scrieciu SŞ, Belton V, Chalabi Z, Mechler R, Puig D. Advancing methodological thinking and practice for development-compatible climate policy planning. Mitigation and Adaptation Strategies for Global Change. 2014;**19**(3):261-288. DOI: 10.1007/s11027-013-9538-z

[21] OECD. Policy Guidance on Integrating Climate Change Adaptation Its Development Cooperation. 2009. DOI: 10.1787/9789264054950-en

[22] Verbruggen A. Revocability and reversibility in societal decision-making. Ecological Economics. 2013;**85**:20-27. DOI: 10.1016/j.ecolecon.2012.10.011

[23] Mediation Toolbox [Internet]. 2013. Available from: http://www.mediation-project.eu/platform/tbox/cba.html [Accessed: 20-06-2018]

[24] Porthin M, Rosqvist T, Perrels A, Molarius R. Multi-criteria decision analysis in adaptation decision-making: A flood case study in Finland. Regional Environmental Change. 2013;**13**(6):1171-1180. DOI: 10.1007/s10113-013-0423-9

[25] Molarius R, Perrels A, Porthin M, Rosqvist T. Testing a Flood Protection Case by Means of a Group Decision Support System, VATT-keskustelualoitteita 449, Finnish Government Institute for Economic Research. 2008

[26] Toth FL, Hizsnyik E. Managing the inconceivable: Participatory assessments of impacts and responses to extreme climate change. Climatic Change. 2008;**91**(1-2):81-101. DOI: 10.1007/s10584-008-9425-x

[27] SFS-ISO 31000:2018. Risk Management. Guidelines. Helsinki: Finnish Standards Association; 2018. p. 39

[28] Rouhianen V. Modelling of accident sequences. In: Rouhiainen V, Suokas J, editors. Quality Management of Safety and Risk Analysis. Amsterdam: Elsevier; 1993. ISBN: 0-444-89864-6

[29] Andrews JD, Dunnett SJ. Event-tree analysis using binary decision diagrams. IEEE Transactions on Reliability. 2000;**49**(2):230-238. DOI: 10.1109/24.877343

[30] Rosqvist T, Molarius R, Virta H, Perrels A. Event tree analysis for flood protection—An exploratory study in Finland. Reliability Engineering and System Safety. 2013;**112**:1-7. DOI: 10.1016/j.ress.2012.11.013

[31] Åström H, Hansen PF, Garre L, Arnbjerg-Nielsen K. An influence diagram for urban flood risk assessment through pluvial flood hazards under non-stationary conditions. Journal of Water and Climate Change. 2014;**5**(3):276-286. DOI: 10.2166/wcc.2014.103

[32] Haer T, Wouter Botzen WJ, Zavala-Hidalgo J, Cusell C, Ward PJ. Economic evaluation of climate risk adaptation strategies: Cost-benefit analysis of flood protection in tabasco, Mexico. Atmosfera. 2017;**30**(2):101-120. DOI: 10.20937/ATM.2017.30.02.03

[33] Zhou Q, Mikkelsen PS, Halsnæs K, Arnbjerg-Nielsen K. Framework for economic pluvial flood risk assessment considering climate change effects and adaptation benefits. Journal of Hydrology. 2012;**414-415**:539-549. DOI: 10.1016/j.jhydrol.2011.11.031

[34] Matko M, Golobic M, Kontic B. Integration of extreme weather event risk assessment into spatial planning of electric power infrastructure. Urbani Izziv. 2016;**27**(1):95-112. DOI: 10.5379/urbani-izziv-en-2016-27-01-001

[35] Ryan PC, Stewart MG. Cost-benefit analysis of climate change adaptation for power pole networks. Climatic Change. 2017;**143**(3):519-533. DOI: 10.1007/s10584-017-2000-6

[36] Kaspersen PS, Halsnæs K. Integrated climate change risk assessment: A practical application for urban flooding during extreme precipitation. Climate Services. 2017;**6**:55-64. DOI: 10.1016/j.cliser.2017.06.012

[37] Papathoma-Köhle M, Promper C, Glade T. A common methodology for risk assessment and mapping of climate change related hazards—Implications for climate change adaptation policies. Climate. 2016;**4**(1):8. DOI: 10.3390/cli4010008

[38] Zanuttigh B, Simcic D, Bagli S, Bozzeda F, Pertrantoni L, Zagonari F, et al. THESEUS decision support system for coastal risk management. Coastal Engineering. 2014;**87**:218-239. DOI: 10.1016/j.coastaleng.2013.11.013

[39] Rumson AG, Hallett SH. Opening up the coast. Ocean and Coastal Management. 2018;**160**:133-145. DOI: 10.1016/j.ocecoaman.2018.04.015

[40] Lieske DJ. Coping with climate change: The role of spatial decision support tools in facilitating community adaptation. Environmental Modelling & Software. 2015;**68**:98-109. DOI: 10.1016/j.envsoft.2015.02.005

[41] Ferreira O, Viavattene C, Jiménez JA, Bolle A, das Neves L, Plomaritis TA, et al. Coastal Engineering. 2018;**134**:241-253. DOI: 10.1016/j.coastaleng.2017.10.005

[42] Jun K, Chung E, Kim Y, Kim Y. A fuzzy multi-criteria approach to flood risk vulnerability in South Korea by considering climate change impacts. Expert Systems with Applications. 2013;**40**(4):1003-1013. DOI: 10.1016/j.eswa.2012.08.013

[43] Jeong S, An Y-Y. Climate change risk assessment method for electrical facility. In: International Conference on Information and Communication Technology Convergence (ICTC 2016); November 2016. DOI: 10.1109/ICTC.2016.7763464

[44] Wilby RL, Dessai S. Robust adaptation to climate change. Weather. 2010;**65**(7):180-185. DOI: 10.1002/wea.543

[45] Lempert RJ, Groves DG. Identifying and evaluating robust adaptive policy responses to climate change for water management agencies in the American west. Technological Forecasting and Social Change. 2010;**77**(6):960-974. DOI: 10.1016/j.techfore.2010.04.007

[46] Shortridge J, Guikema S, Zaitchik B. Robust decision making in data scarce contexts: Addressing data and model limitations for infrastructure planning under transient climate change. Climatic Change. 2017;**140**(2):323-337. DOI: 10.1007/s10584-016-1845-4

[47] Rød JK, Opach T, Neset T-S. Three core activities toward a relevant integrated vulnerability assessment: Validate, visualize, and negotiate. Journal of Risk Research. 2015;**18**(7):877-895. DOI: 10.1080/13669877.2014.923027

[48] Henriksen HJ, Roberts MJ, van der Keur P, Harjanne A, Egilson D, Alfonso L. Participatory early warning and monitoring systems: A Nordic framework for web-based flood risk management. International Journal of Disaster Risk Reduction. 2018. DOI: 10.1016/j.ijdrr.2018.01.038

[49] van Aalst MK, Cannon T, Burton I. Community level adaptation to climate change: The potential role of participatory community risk assessment. Global Environmental Change. 2008;**18**:165-179. DOI: 10.1016/j.gloenvcha.2007.06.002

[50] Lépy E, Heikkinen HI, Karjalainen TP, Tervo-Kankare K, Kauppila P, Suopajärvi T, et al. Multidisciplinary and participatory approach for assessing local vulnerability of tourism industry to climate change. Scandinavian Journal of Hospitality and Tourism. 2014;**14**(1):41-59. DOI: 10.1080/15022250.2014.886373

[51] Juhola S, Driscoll P, Mendler de Suarez J, Suarez P. Social strategy games in communicating trade-offs between mitigation and adaptation in cities. Urban Climate. 2013;**4**:102-116. DOI: 10.1016/j.uclim.2013.04.003

[52] Neuvonen M, Sievänen T, Fronzek S, Lahtinen I, Veijalainen N, Carter TR. Vulnerability of cross-country skiing to climate change in Finland—An interactive mapping tool. Journal of Outdoor Recreation and Tourism. 2015;**11**:64-79. DOI: 10.1016/j.jort.2015.06.010

[53] Armaroli C, Duo E. Validation of the coastal storm risk assessment framework along the Emilia-Romagna coast. Coastal Engineering. 2018;**134**:159-167. DOI: 10.1016/j.coastaleng.2017.08.014

[54] Rizzi J, Torresan S, Zabeo A, Critto A, Tosoni A, Tomasin A, et al. Assessing storm surge risk under future sea-level rise scenarios: A case study in the North Adriatic coast. Journal of Coastal Conservation. 2017;**21**:453. DOI: 10.1007/s11852-017-0517-5

[55] Bles T, Bessembinder J, Chevreuil M, Danielsson P, Falemo S, Venmans A, et al. Climate change risk assessments and adaptation for roads—Results of the ROADAPT project. Transportation Research Procedia. 2016;**14**:58-67. DOI: 10.1016/j.trpro.2016.05.041

[56] Tagg A, Räikkönen M, Mäki K, Roca Collell M. Impact of extreme weather on critical infrastructure: The EU-INTACT risk framework. In: E3S Web of Conferences e3sconf/201 FLOOD risk 2016—3rd European Conference on Flood Risk Management. 2016

[57] Molarius R, Könönen V, Leviäkangas P, Rönty J, Hietajärvi A-M, Oiva K. The extreme weather risk indicators (EWRI) for the European transport system. Natural Hazards Springer. 2014;**72**(1):189-210. DOI: 10.1007/s11069-013-0650-x

[58] Hallegatte S, Ranger N, Mestre O, Dumas P, Corfee-Morlot J, Herweijer C, et al. Assessing climate change impacts, sea level rise and storm surge risk in port cities: A case study on Copenhagen. Climatic Change. 2011;**104**:113-137. DOI: 10.1007/s10584-010-9978-3

[59] Wade SD, Rance J, Reynard N. The UK climate change risk assessment 2012: Assessing the impacts on water resources to inform policy makers. Water Resources Management. 2013;**27**(4):1085-1109. DOI: 10.1007/s11269-012-0205-z

[60] Monbaliu J, Chen Z, Felts D, Ge J, Hissel F, Kappenberg J, et al. Risk assessment of estuaries under climate change: Lessons from Western Europe. Coastal Engineering. 2014;**87**:32-49. DOI: 10.1016/j.coastaleng.2014.01.001

[61] Szewranski S, Chruściński J, Kazak J, Świąder M, Tokarczyk-Dorociak K, Żmuda R. Pluvial flood risk assessment tool (PFRA) for rainwater management and adaptation to climate change in newly urbanised. Water. 2018;**10**:386. DOI: 10.3390/w10040386

[62] Straatsma MW, Vermeulen PTM, Kuijper MJM, Bonte M, Niele FGM, Bierkens MFP. Rapid screening of operational freshwater availability using global models. Water Resources Management. 2016;**30**:3013-3026. DOI: 10.1007/s11269-016-1327-5

[63] Iyalomhe F, Rizzi J, Pasini S, Torressan S, Critto A, Marcomini A. Regional risk assessment for climate change impacts on coastal aquifers. The Science of the Total Environment. 2015;**537**:100-114. DOI: 10.1016/j.scitotenv.2015.06.111

[64] Veijalainen N. Estimation of climate change impacts on hydrology and floods in Finland [Ph.D. thesis]. Helsinki, Finland: Aalto University; 2012

[65] Knight PJ, Prime T, Brown JM, Morrissey K, Plater AJ. Application of flood risk modelling in a web-based geospatial decision support tool for coastal adaptation to climate change. Natural Hazards and Earth System Sciences. 2015;**15**:1457-1471. DOI: 10.5194/nhess-15-1457-2015

[66] Molarius R, Räikkönen M, Forssén K, Mäki K. Enhancing the resilience of electricity networks by multi-stakeholder risk assessment: The case study of adverse winter weather in Finland. Journal of Extreme Events World Scientific. 2016;**3**(4). DOI: 10.1142/S2345737616500160

[67] Pasini S, Torressan S, Rizzi J, Zabeo A, Critto A, Marcomini A. Climate change impact assessment in Veneto and Friuli Plain groundwater. Part II: A spatially resolved regional risk assessment. Science of the Total Environment. 2012;**440**:219-235. DOI: 10.1016/j.scitotenv.2012.06.096

Chapter 4

Statistical Methodology for Evaluating Process-Based Climate Models

Firdos Khan and Jürgen Pilz

Additional information is available at the end of the chapter

http://dx.doi.org/10.5772/intechopen.80984

Abstract

In climatology, there are mainly two types of models used, that is, global circulation/climate models (GCMs) and regional climate models (RCMs). GCMs can be run for the whole globe, while RCMs can be run only for a part of the globe. In this chapter, we provided a general statistical methodology for evaluating process-based (GCM or RCM) climate models. To bridge observed and simulated data sets, statistical bias correction was implemented. A meta-analysis technique is used for selecting a model or scenarios, which have good performance compared to others. For model selection and ensemble projection, Bayesian model averaging (BMA) is used. Posterior inclusion probability (PIP) is used as model selection criterion. Our analysis concluded with a list of best models for maximum, minimum temperature, and precipitation where the rank of the selected models is not the same for the listed three variables. The outputs of BMA closely followed the pattern of observed data; however, it underestimated the variability. To overcome this issue, 90% prediction interval was calculated, and it showed that almost all the observed data are within these intervals. The results of Taylor diagram show that the BMA projected data are better than the individual GCMs' outputs.

Keywords: bias correction, climate change, meta analysis, model selection, posterior inclusion probability

1. Introduction

This chapter is basically about statistical evaluation of climate models; however, prior to model's evaluation, it is important to highlight briefly about climate models and their types. In the literature, climate models are also known as process-based models as these models work closely to the physical process of the climate of our planet. Broadly speaking, climate models can be divided into two types, global circulation/climate model (GCM) and regional climate model (RCM). As the name of these models suggests, GCM can be run for the whole globe, while RCM can be run for a particular location of interest. Due to different uncertainties, we need to evaluate these models before using their outputs for further analysis. Evaluation of a climate model means assessing a model's performance so as to articulate the grounds on which a model can be declared good enough for its anticipated use. Model's evaluation is an important step in climate change assessment and impact assessment studies. It provides guide lines to choose the best models or scenarios for further analysis. For evaluation of climate models, we need observational data and historical simulated data by using climate models for the same variables and for the same time period. This chapter presents a combination of classical and Bayesian statistical approaches for the evaluation of climate models. This chapter is structured as: right after this brief introduction, climate models in general and GCM and RCM in particular and their evaluation are briefly explained, and methodology is discussed in Section 2, Section 3 is reserved for results and discussion, and Section 4 comprised of summary and conclusion.

1.1. Climate model

A climate model is a complex system of mathematical equations which represents the physical process among various components which contribute to the climate of our globe. To run a GCM, the globe is divided into a number of grid boxes with horizontal resolution (latitude and longitude) and vertical resolution (height or pressure). The climatology is solved in each grid after providing the initial conditions to the main deriving climate variables like temperature, wind speed, humidity, pressure, etc. The brief details about GCMs and RCMs are presented in the subsequent sections, for more details about GCMs, we refer to [1].

1.2. Global climate model

GCMs are the most modern and sophisticated tools available to provide basic information about climate globally. These models comprise complex mathematical equations which represent the physical process of atmosphere, ocean, cryosphere, and land surface. GCMs are run for the whole globe, and due to the complex system, it takes long time in simulations; however, super computers can be used to make their performances more efficient. According to [1], a GCM is "Numerical models, representing physical processes in the atmosphere, ocean, cryosphere and land surface, are the most advanced tools currently available for simulating the response of the global climate system to increasing greenhouse gas concentrations." The process of simulating climate systems by using climate models is called dynamical modeling or dynamical downscaling. The nesting of dynamical modeling and simulation of climate systems is presented in **Figure 1** using GCM and RCM, while a list of some popular GCMs is presented in **Table 1** along with some basic information about each model.

1.3. Regional climate model

RCMs are also process-based climate models and comprise complex mathematical equations like GCMs; however, these models can be run for a particular location of interest. As GCMs can be run for the whole globe, therefore, the grid size is coarser, and the information in the

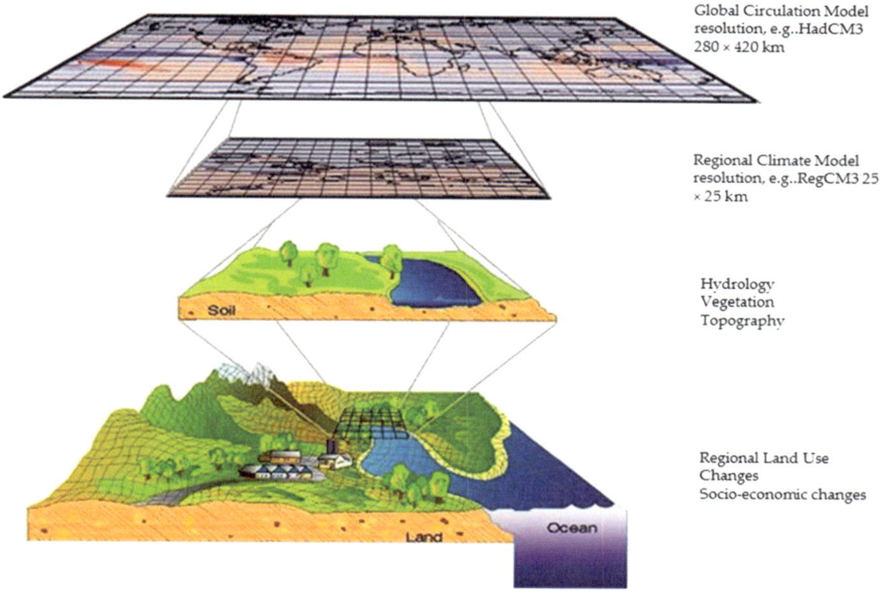

Figure 1. The chain of dynamical downscaling, global climate modeling, and then regional climate modeling to get climate information at higher resolution from coarser resolution.

S. No.	Name	Resolution (Degree*)	Institute/Center
1	CanESM2	2.7906 × 2.8125	Canadian Centre for Climate Modeling and Analysis
2	CCSM4	1.250 × 0.942	National Center for Atmospheric Research
3	CESM1-CAM5	1.250 × 0.942	National Center for Atmospheric Research
4	CMCC-CMS	1.875 × 1.865	Centro Euro-Mediterraneo per I Cambiamenti Climatici
5	CNRM-CM5	1.406 × 1.401	Centre National de Recherches Meteorologiques
6	EC-EARTH	1.1215 × 1.125	Irish Centre for High-End Computing (ICHEC), European Consortium
7	GFDL-ESM2G	2.500 × 2.023	Geophysical Fluid Dynamics Laboratory
8	GFDL-ESM2M	2.500 × 2.023	Geophysical Fluid Dynamics Laboratory
9	INM-CM4	2.000 × 1.500	Institute for Numerical Mathematics
10	MIROC-ESM-CHEM	2.7906 × 2.8125	National Institute for Environmental Studies, The University of Tokyo
11	MPI-ESM-LR	1.875 × 1.865	Max Planck Institute for Meteorology (MPI-M)
12	MPI-ESM-MR	1.875 × 1.865	Max Planck Institute for Meteorology (MPI-M)
13	NorESM1-M	2.500 × 1.895	Norwegian Climate Centre

*1 degree is approximately equal to 111.32 km.

Table 1. Details about GCMs used in this study with their resolutions and other brief information.

form of output of GCMs is at lower resolution. For impact assessment studies like impact of climate change on water resources, agriculture, urban planning, etc., we need information at higher resolution. Toward this end, we need to do regional climate modeling which provides climatic information at higher resolutions. Due to the rapid development in computational technology, the modern RCMs can be run with a resolution of 10 km or even higher resolution.

1.4. Evaluation of climate models

Evaluation of the process-based climate models is an important step in climate change assessment and impact assessment studies. Different methods can be used for this purpose; however, we presented a combination of advanced statistical methods for model evaluation including classical and Bayesian approaches. Evaluation of process-based climate model will give guide lines about model or scenario selection to researchers in climate changes assessment and studies related to the impact assessment of climate change in different areas. This will help researchers to use specified and representative models rather than randomly selected model and to get realistic results. Statistical bias correction is important to reduce the gap between observed and model's simulated data and considers initial step toward climate change assessment. Meta analysis is used for scenario analysis, which assigns higher weights on the basis of precision of a particular scenario. For model (GCM in our case), we used Bayesian model averaging (BMA) technique to choose a set GCMs performing better than others. On the basis of chosen GCMs, ensemble projections were calculated using BMA technique and further evaluated using Taylor diagram along with individual GCMs' outputs. Further details are presented in methodology section about each method and examples with discussion in results and discussion section.

1.5. Objectives of this study

This study aims to present statistical methods including frequentist as well as Bayesian which can be used for the evaluation of process-based models with detailed examples and discussion using the real world data.

2. Data and methodology

2.1. Data

Two types of data sets have been used in this research, observed and climate model's simulated data. Observational data were acquired from Pakistan Meteorological Department (PMD) on daily frequency for three climate variables including maximum, minimum temperature, and precipitation for different locations of Pakistan. RCM's simulated data were collected from COordinated Regional climate Downscaling Experiment (CORDEX), Met. Office of the United Kingdom (UK), and Global Change Impact Studies Centre (GCISC), Pakistan. For evaluation purpose, we used data for the baseline time period, which is 1960–1990 for both observed and simulated data, but for calculating ensemble projections and their evaluation, the baseline is considered as 1975–2005.

2.2. Statistical bias correction

According to [2], all models are wrong but some of them are useful. Climate models provide useful information; however, there are various sources of uncertainties which have influence on the outputs of these models [3, 4]. To bridge the difference between observed and model's simulated data, we need to utilize statistical methods. In order to carry out statistical bias correction, different statistical methods were developed starting from simple to most sophisticated ones. For detailed literature about statistical bias correction methods, we refer to [5, 6]. We present the methodology of latest developed methods by [7], which preserve trend and climate extremes in future climate model's simulations called quantile delta mapping (QDM).

A four step methodology is required to implement the QDM method, starting from the cumulative distribution function (CDF) of model projected series $Y_{m,f}$. We assume that f, h, m, and o stand for future, historical, model, and observed data, respectively. Further, F and Y represent CDF of the data and original data, respectively.

$$F_{m,f}(y(t)) = P\left(Y_{m,f}(t) \leq y(t)\right), F_{m,f}(t) \in [0, 1] \tag{1}$$

To proceed to the second step, we need to find the relative change using the ratio of the inverse CDF of model predicted data applied to the CDF of model predicted data and the inverse CDF of historical observed data applied to model predicted data. Mathematically, this can be written by Eq. (2).

$$\Delta_m(y(t)) = \frac{F_{m,f}^{-1}\left(F_{m,f}(y(t))\right)}{F_{m,h}^{-1}\left(F_{m,f}(y(t))\right)} = \frac{y(t)}{F_{m,h}^{-1}\left(F_{m,f}(y(t))\right)} \tag{2}$$

The quantiles of model's predicted data $F_{m,f}(y(t))$ can now be bias corrected by implementing the inverse CDF estimated from historical observational data.

$$\hat{Y}_{o,m}(t) = F_{o,h}^{-1}\left(F_{m,f}(y(t))\right) \tag{3}$$

Finally, the bias corrected future projections can be obtained by applying the relative changes to the historical bias corrected data presented in Eq. (4).

$$\hat{Y}_{m,f}(t) = \hat{Y}_{o,m}(t).\Delta_m(y(t)) \tag{4}$$

$\hat{Y}_{m,f}(t)$ is the future model's bias corrected data which can be used now for further analysis. To preserve absolute changes in the data, Eqs. (2) and (4) can be applied additively rather than multiplicatively [7]. The multivariate counterpart of the method presented in [7] is available and can be found in [8]. One advantage of multivariate quantile mapping bias correction (presented in [8]) is that it preserves spatial dependence structures between climate variables when we are applying this method to more than one variable simultaneously. This is especially important when we are dealing with impact assessment studies like hydrological modeling, agricultural production, etc. under the changing climate.

2.3. Meta analysis

Scenario's or model's assessment is an essential part of climate change analysis as it provides valuable information about a particular scenario or model. Meta analysis is a statistical method which can be used for this purpose; however, it is also a useful technique to produce a combined estimate of projections from individual model outputs or different scenarios. It gives weight to each study on the basis of its precision and, consequently, provides confidence in future projections which have higher precision. Usually, researchers prefer models, scenarios, studies, laboratories' outputs, etc., which have higher weights than those with lower weights. In order to accomplish the evaluation of models or scenarios using meta analysis, the three step methodology is explained briefly in the following subsections.

2.3.1. *Selection of the model*

There are two basic models to perform a meta analysis: the fixed effect model (FEM) and the random effect model (REM) [9]. The FEM assumes that all the studies included in the meta analysis come from a single identical population or share a common effect (mean or average), while a REM assumes that the effects of the studies included in the meta analysis form a random sample from a population following a specified distribution. The observed effects in the FEM and REM are mathematically presented in Eqs. (5) and (6), respectively. Suppose we have k studies, and let θ denote the (true) intervention effect in the population, which we would like to estimate. Further, let θ_k denote the kth study effect, and ζ_k the random effect in this study; $k = 1, 2, \ldots, K$.

$$\theta_k = \theta + \varepsilon_k,\ \varepsilon_k \sim N(0, v_k^2) \tag{5}$$

$$\theta_k = \theta + \zeta_k + \varepsilon_k,\ \zeta_k \sim N(0, \tau^2) \tag{6}$$

Here, ε_k describes the variation within the kth study, and the random effects ζ_k reflect the variations between the considered studies. FEM is a special case of REM when the variations between studies are equal to zero:

$$\zeta_1 = \zeta_2 = \ldots = \zeta_K = 0, \tag{7}$$

then the random effect model reduces to the fixed effect model. Model selection is mainly based on the nature and objectives of the study [9–11]. This is an important step because the remaining two steps depend on model selected. Therefore, model selection must be made carefully. This chapter presented both models for explanation, and in results and discussion section, we present an example to make clear the differences between these two models.

2.3.2. *Weighting schemes for parameter estimation*

Different weighting schemes are available for the estimation of the effect size in meta analysis; however, it depends on the nature of the study to choose one of them [9]. We proceed with the so-called inverse-variance weighting technique for quantifying the effect size in our analysis. For details about different weighting schemes, we refer to [10]. According to [9], all the available schemes are efficient because they assign higher weights to more precise studies. In case of a fixed-effect model, the weights are calculated by Eq. (8).

$$\omega_k = \frac{1}{v_k^2} \tag{8}$$

where ω_k and v_k^2 are the weight and variance, respectively, of the kth study. In a random effect model, the weights are calculated by Eq. (9).

$$\begin{aligned} \omega_k^* &= \frac{1}{v_k^*} \\ v_k^* &= v_k^2 + \tau^2 \end{aligned} \tag{9}$$

where ω_k^* is the weight for kth study, and v_k^* is the combined variance of within-study and between-studies (τ^2). As we had already discussed, the weights for estimating the effect size depend on the model chosen in the model specification stage.

2.3.3. *Estimation of parameters*

The next step is to estimate the unknown parameters of the specified model by incorporating the weighted least squares method given by Eq. (10).

$$\begin{aligned} \theta_c &= \frac{\sum_{k=1}^{K} W_k.\theta_k}{\sum_{k=1}^{K} W_k} \\ W_k &= \frac{1}{var(\theta_k)} \end{aligned} \tag{10}$$

The $(1-\alpha) \times 100\%$ confidence interval of the combined estimator is given by

$$\theta_c \pm Z_{\left(1-\frac{\alpha}{2}\right)} \times SE(\theta_c)$$

where θ_c is the combined size effect, $SE(\theta_c)$ is the standard error, and $Z_{\left(1-\frac{\alpha}{2}\right)}$ is the $\left(1-\frac{\alpha}{2}\right)$-quantile of the standard normal distribution.

2.4. Model (GCM) selection, ensemble projections, and their evaluation

In this part, Bayesian model averaging (BMA) is used for two purposes, that is, model selection and producing ensemble projections by using the outputs of selected GCMs and finally the evaluation of models' outputs. **Table 1** lists some popular GCMs with brief details about their resolution, and the institutes where each model was developed. The outputs of these models are used in the subsequent sections of this chapter. However, before embarking on this journey, it is important to discuss briefly the concept of posterior probability which is at the core of the Bayesian approach. Bayesian model averaging is discussed afterwards.

2.4.1. Posterior probability

Bayes' theorem states that the posterior probability of jth model, $p(M_j \mid D)$, is calculated as the likelihood of observed data given j^{th} model, $p(D \mid M_j)$, multiplied by the prior probability of the jth model, and divided by the probability of having the current observation realization, $p(D)$. The posterior probability is thus calculated as follows:

$$P(M_j \mid D) = \frac{p(D \mid M_j).p(M_j)}{p(D)} \tag{11}$$

In Eq. (11), $p(D)$ is used as a normalizing constant given in Eq. (12), and hence the Bayes' rule can be simply stated as in Eq. (13).

$$p(D) = \sum_{j=0}^{s} p(D \mid M_j).p(M_j) \tag{12}$$

$$p(M_j \mid D) \propto p(D \mid M_j).p(M_j) \tag{13}$$

The prior distribution of a model shows the probability allocated to a statistical model. In this study, we have $M_j; j = 0, 1, 2, \ldots, s = 2^k\text{-}1$ possible statistical models. The likelihood of observation represents the probability of getting the current model realization. The posterior probability of a model represents the probability of the model to realize the current model given observations. Different choices for the prior are available; however, the users can also implement their own customized priors for their analysis. In case of using a uniform prior distribution (i.e., $p(M_j) \propto 1/2^k$), assigning equal prior weight to all models then the posterior model probability can be expressed by Eq. (14).

$$p(M_j \mid D) \propto p(D \mid M_j) \tag{14}$$

Eq. (14) shows that in this case the posterior probability of a model is only determined by the likelihood of observational data. Likelihood of a model reflects the ability to reproduce a given system of observed data. Different likelihood functions have been proposed to calculate the likelihood, $p(D \mid M_j)$, for example, see [13–15, 17]. A Gaussian likelihood proposed by [16] is used in this chapter.

2.4.2. GCM selection

The rapid developments in the computational technology make it practicable to run complex process-based climate models for simulating complex climate systems of our planet;

however, the output from a single model still may have uncertainties [18]. There has been a number of GCMs developed to project the future global climate change and use their output for impact assessment studies in different areas [1]. Due to different parameterization schemes of GCMs, internal atmospheric variability [19], and uncertainties in input data, different GCMs may produce quite different results. Therefore, it is important to consider more GCMs instead of relying on a single GCM. Regression models can be used to estimate the observed climate by using outputs from different GCMs as covariates. In a regression model context, the problem of uncertainty modeling has been raised by Raftery et al. [20]. In such models, covariate (GCM here) selection is a basic part to build a valid regression model, and the objective is then to find the "best" model for the response variable and a given set of predictors. The first problem to solve is which covariates should be included in the model and how important are they? Suppose we have a response variable Y and set of covariates, $X_1, \ldots, X_k$, and E represents the expected value, then there are 2^k different linear regression models expressing the relationship between the response variable and the potential predictors as follows:

$$\begin{gathered}M_0{:}EY = \beta_0, M_1{:}EY = \beta_0 + \beta_1 X_1, \ldots, M_k{:}EY = \beta_0 + \beta_k X_k \\ M_{k+1}{:}EY = \beta_0 + \beta_1 X_1 + \beta_2 X_2, \ldots, M_{k+l}{:}EY = \beta_0 + \beta_{k-1} X_{k-1} + \beta_k X_k; l = k(k-1)/2 \\ M_{k+l+1}{:}EY = \beta_0 + \beta_1 X_1 + \beta_2 X_2 + \beta_3 X_3, \ldots, M_{k+l+m}{:}EY = \beta_0 + \beta_{k-2} X_{k-2} + \beta_{k-1} X_{k-1} + \beta_k X_k, \\ m = \frac{k(k-1)(k-2)}{6} \\ M_{2^k-1}{:}EY = \beta_0 + \beta_1 X_1 + \beta_2 X_2 + \ldots + \beta_k X_k\end{gathered} \tag{15}$$

The same procedures are used here for GCM selection, where GCM now stands for a model M_j; $j = 1, \ldots, m = 13$; and a uniform prior was used as a prior probability distribution for GCMs ($p(GCM) \propto \frac{1}{m}$). The posterior inclusion probability (PIP) is the sum of posterior probabilities of each covariate (GCM) from all possible models included in BMA used as model selection criterion. The PIP has a range between zero and one, where a value close to one means that the GCM closely reproduces the observed data, while a value close to zero means that the corresponding GCM's output does not agree at all with the observed data.

It has a strong and solid background in Bayesian statistics, and there is a rich body of literature on BMA and PIP. GCM selection was made along the following four steps:

1. Run BMA as presented in Eq. (15).
2. Calculate the posterior probability for each GCM included in all possible regression models in step 1.
3. Sum the posterior probabilities for each GCM from all possible models called PIP in step 2.
4. Decide about the models having higher PIPs.

As the criterion is probability; therefore, we prefer models with higher PIPs than those with lower PIPs. Similarly, this procedure can be used for model selection in other areas like hydrology, ecology, forestry, etc.

2.4.3. Ensemble projections

Normally, it is assumed in standard regression modeling that a single model be the true model to examine the response variable given a set covariates, but other probable models could give different outcomes for the same problem at hand. The typical approach, which means conditioning on a single model supposed to be true, nevertheless, it does not take account of model uncertainties. One way is to compute an arithmetic ensemble mean (AEM) as a prediction as this could provide better results than any of the single model's output; however, this approach gives no information about the uncertainty that the predictions have [21]. BMA overcomes this issue by estimating the regression models using all possible combinations of covariates given in Eq. (15) and then builds a weighted average model from all possible models. Thus, it provides probabilistic projections where the weights are the posterior probabilities during the training period, and these are directly tied to the performance of the models [20, 22–24]. The predictive probability density function (PPDF) of BMA of a variable of interest is the weighted average of PDFs of individual forecasts where the weights are the posterior model probabilities [20]. The performance of BMA is considered better in different areas such as ground water modeling, weather forecast, hydrological predictions, and model uncertainty analysis [25–31]. Suppose we have a set of k covariates (different GCMs' outputs in this study), then there are 2^k statistical models $M_0, \ldots, M_s$. Then, the conditional forecast PDF of the variable of interest on the basis of training data D (observational data) is presented in Eq. (16).

$$p(y \mid D) = \sum_{j=0}^{s} p(y \mid M_j, D).p(M_j \mid D) \tag{16}$$

where $p(y \mid M_j)$ is the forecast PDF based on model M_j, and $p(M_j \mid D)$ is the corresponding posterior probability used as a weight; consequently, it reflects how well the model fits the data during the training time period. As the weights are posterior probabilities presented in Eq. (16), therefore $\sum_{j=0}^{s} p(M_j \mid D) = 1$. The posterior mean and variance of PDF in Eq. (16) can be easily calculated and are given in Eqs. (17) and (18), respectively [24].

$$E(y \mid D) = \sum_{j=0}^{s} E(y \mid M_j, D).p(M_j \mid D) = \sum_{j=0}^{s} \mu_j.w_j \tag{17}$$

$$Var(y \mid D) = \sum_{j=0}^{s} \left(\mu_j - \sum_{i=0}^{s} w_i.\mu_i\right)^2 + \sum_{j=0}^{s} w_j.\sigma^2_j \tag{18}$$

$$\text{Predictive variance} = \text{Between Model Variance} + \text{Within Model Variance} \tag{19}$$

In Eq. (18), the predictive variance has two parts, one is the between models variance, and the other one is the within model variance [20]. The variance σ^2_j is associated with the jth model's prediction. The between model variance indicates how the individual model mean predictions deviate from the ensemble prediction, with all contributions to deviations weighted by posterior model weights [26]. The within model variance represents the individual model contributions weighted by the corresponding posterior model probability. Whether to consider only one of them or both depends on the objectives of the study. In the example discussed in results and discussion section, we are not taking into account the second term as the objective is the ensemble assessment of climate change as suggested by [26]. As the prediction mean and variance of the forecasted PDF are available now, the prediction interval can be constructed using Eq. (20).

$$\mu_{pr} \pm z_{\left(1-\frac{\alpha}{2}\right)} \cdot \sqrt{Var_{pr}} \quad (20)$$

Here, μ_{pr}, Var_{pr}, and $z_{\left(1-\frac{\alpha}{2}\right)}$ are ensemble mean, variance, and the $\left(1-\frac{\alpha}{2}\right)$-quantile of the standard normal distribution, respectively. The subscript pr with μ_{pr}, Var_{pr} means that these statistics are about the predicted PDF.

2.4.4. Taylor diagram

Taylor diagrams are rather sophisticated diagrams for graphical evaluation of a system, process or phenomenon. It was invented by [31] in 1996; however, it was published later in 2001 to aid researchers in comparative assessment of the performance of different models. The diagram is used to quantify the degree of correspondence between observational and modeled data sets in term of three statistics, standard deviation (SD), root mean square error (RMSE), and correlation coefficient (CC). We used the R software system to create the Taylor diagrams (a) for maximum temperature, (b) for minimum temperature, and (c) for precipitation presented in **Figure 6**. The data on all these variables are taken from northern Pakistan.

The interpretation of Taylor diagrams is straight forward; however, sometimes it feels tricky and needs basic understanding of statistics. The model's performance is considered better if the modeled and observed data have strong correlation, and the modeled data have low RMSE and have closer standard deviation to that of the observational data.

3. Results and discussion

This section provides analysis and results about the methodology presented in Section 2. Each section presented in methodology section is explained with examples and detailed discussion.

3.1. Statistical bias correction

Figure 2 presents the results about evaluation of statistical bias correction methods applied to temperature and rainfall data taken from Northern Pakistan. In **Figure 2**, observed represents observed data, Sim-baseline is model's simulated data for the base line period (1960–1990), and Sim-BC stands for model's simulated data after the application of statistical bias correction techniques. For comparison, we keep the time duration of both data sets (simulated and observed) same (1960–1990). It can be seen from **Figure 2** that there are marked differences between observed and regional climate model's simulated data for temperature and precipitation in Northern Pakistan. The left panel displays maximum temperature, while the right panel is about precipitation. In both parts, original simulated results deviated from observational data; however, after the application of statistical bias correction techniques, the

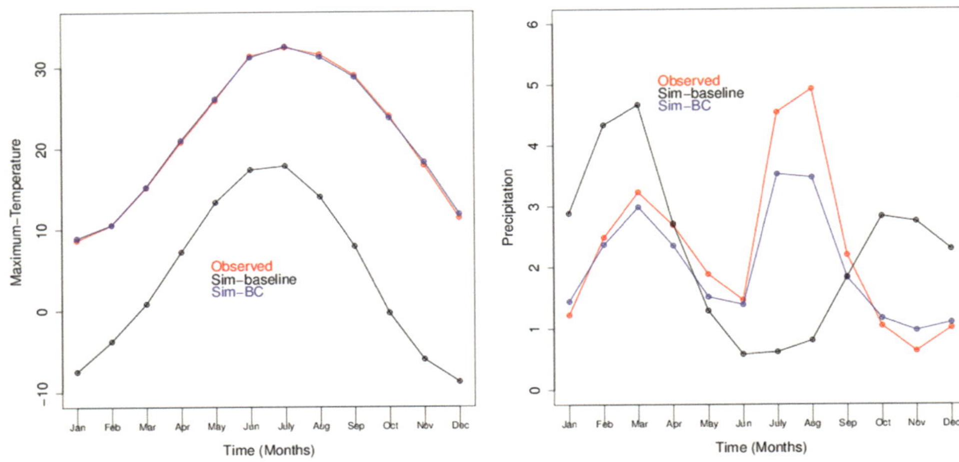

Figure 2. Comparison of observed, simulated, and bias corrected maximum temperature and precipitation for Northern Pakistan. In both parts, the red, black, and blue colors show observed, model simulated, and bias corrected data.

differences are reduced, and the pattern is followed in a better way, particularly for precipitation, where we have quite a large difference between observed and simulated precipitation data. This example is about 30 years averaged data for each month; however, these techniques can be applied to data on other frequencies like daily, hourly, etc.

3.2. Meta analysis

In **Figure 3**, climate change scenarios were analyzed, and the same procedures can be performed for model selection depending on the researcher's objective. In **Figure 3**, on the left side under the heading, studies 1, 2, 3, and 4 represent the A2, B2, RCP4.5, and RCP8.5 scenarios, respectively. The former two are chosen from the fourth assessment report (AR4), while the latter are the two scenarios stem from the fifth assessment report (AR5) of the Intergovernmental Panel on Climate Change (IPCC). In this study, scenario analysis was performed for the mean difference between baseline and future time period. The subheadings total, mean, and SD under experimental and control stand for total number of observation included in a particular scenario, mean value of each scenario, and standard deviation of each scenario, respectively. Experimental and control stand for the baseline period and future time period, respectively. The thick black vertical line shows no difference between the mean values of experimental and control periods. The dotted vertical line shows the combined mean difference. Against study 1, there is outcome effect mean difference for scenario 1, similarly for other scenarios. The length of the line on each box shows the width of the confidence interval, and the size of the box (square shape) shows the weight assigned to a particular study. As the weights are assigned on the basis of precision of a scenario (in our case), therefore, the scenarios receiving higher weights have less variance and consequently exhibit shorter confidence intervals. The bigger the square box the higher the weights assigned to a particular scenario. The two diamond shapes represent combined mean differences where the upper one is for FEM, while the lower one is for REM. The width of the diamond shows the

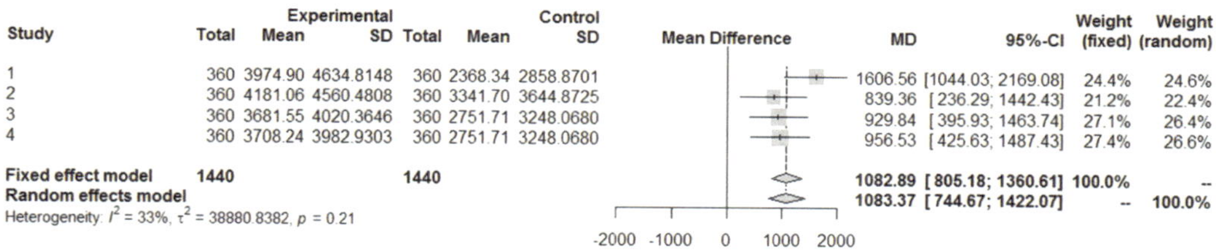

Figure 3. Here, studies 1, 2, 3, and 4 represent A2, B2, RCP4.5, and RCP8.5 scenarios, respectively. Meta analysis was conducted for mean difference between experimental and control time periods.

confidence interval under the selected model, and it is obvious that the width of the diamond for REM is larger than that of FEM. The reason for this difference is that REM also considered the variation between studies (Eq. (9)), while a FEM does not consider this variation (Eq. (8)). Bear in mind that two R packages "metafor" and "meta" have been used for this analysis. For details about these packages, we refer to [11, 12], respectively.

3.3. GCM selection and ensemble projection

3.3.1. GCM selection

Figure 4 presents the output of model selection results for 13 different GCMs where the criterion of model selection is PIP. To our knowledge, this technique is used for the first time in atmospheric sciences for the purpose of GCMs selection. The model selection procedure was run multiple times (six times here) with small changes in sample size; however, no significant changes have been noted in the end results. In **Figure 4**, the results are presented only for maximum temperature; nevertheless, the same procedures can be used for other variables.

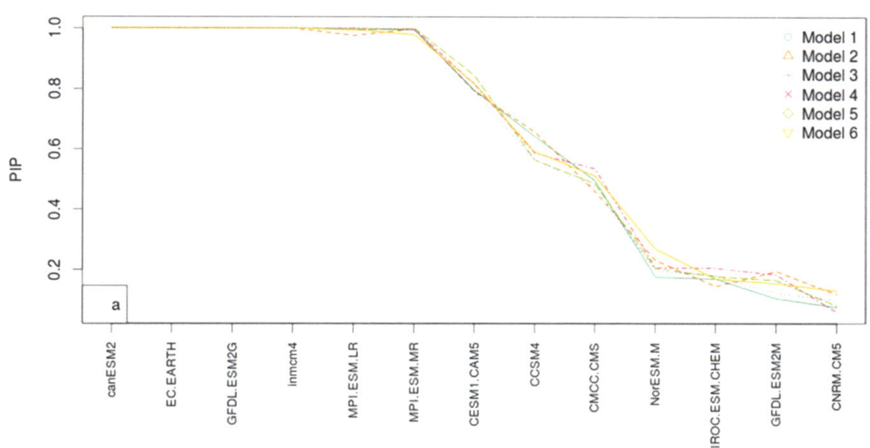

Figure 4. GCM selection among competing models for maximum temperature using posterior inclusion probability (PIP). PIPs are given on the vertical axis, and models (GCMs) are indexed on the horizontal axis. Model 1-model 6 stand for different runs of BMA with different sample sizes.

For maximum temperature, top five selected GCMs are canESM2, CESM-CAM5, EC-EARTH, GFDL-ESM2G, and INM-CM4 in ascending order. It was investigated that different variables (minimum temperature and precipitation in our case) have different lists of top GCM; however, they shared some models in the top list. The list of top five models has maximum PIPs and can be used for further analysis rather than to use all GCMs.

3.3.2. *Ensemble projections*

Figure 5 elaborates the results of BMAs' outputs calculated from selected GCMs compared with observational data sets for three key climate variables, maximum temperature, minimum temperature, and precipitation for the duration of 30 years (1975–2005) considered as baseline period here. As BMA is a regression-based approach, it estimates the mean value and underestimates variation. To cope with this issue, 90% prediction intervals were calculated for each variable and plotted. In **Figure 5**, red, blue, and deep gray colors represent observational, BMAs' outputs, and 90% prediction intervals for each variable. From the upper panel, it is difficult to get a clear conclusion as it is for 30-year data; however, from the bottom panel, clear information can be inferred. We can see from parts d (maximum temperature) and e (minimum temperature) that BMAs' outputs follow the pattern nicely; however, it does not capture well the variability in both cases. The 90% prediction intervals cover almost all the variation for both variables. For precipitation, we calculated upper 90% prediction intervals, and it can be seen that it covers the observational values; however, quite a few values are still outside because precipitation is a more complex phenomenon than the temperature. Similarly, we can calculate ensemble projections from various models or climate change scenarios in hydrology, agriculture, ecology, etc. to address the uncertainty while relying on the single model's output.

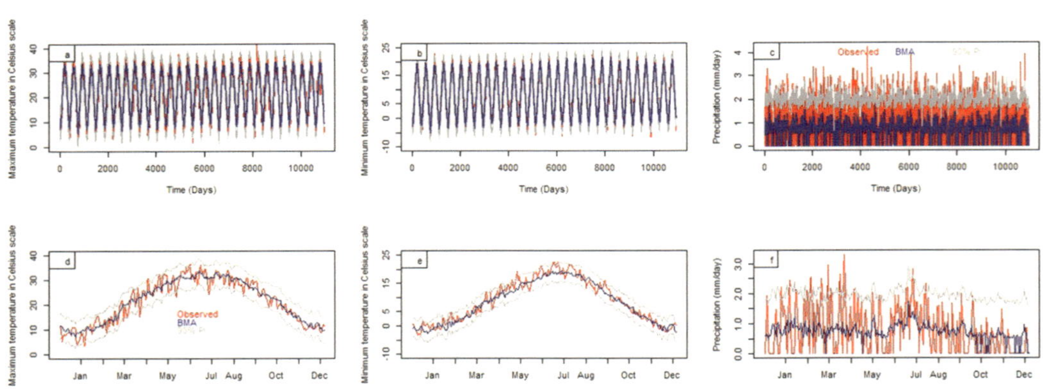

Figure 5. The upper panel demonstrates the evaluation of BMAs' outputs from selected GCMs for a period of 30 years (1975–2005) (a) maximum temperature, (b) minimum temperature, and (c) precipitation. Red, blue, and deep gray colors represent observed, BMA, and 90% prediction interval for each variable. The bottom panel demonstrates the same as in the upper panel for just 1 year.

3.4. Taylor diagram

The green lines inside the graph (**Figure 6**) show RMSE between observed and modeled data sets, while the blue lines show standard deviation of each data set. The straight lines inside the graph show the correlation coefficient between modeled and observational data sets. The dots on the horizontal lines represent observed data in each part of **Figure 6**. Look at part (a) of **Figure 6** which is about maximum temperature and the purpose is to assess which model's simulated data is best as compared to other models. We calculated ensemble projections by using the outputs from all selected GCMs by using BMA and plotted them in each part of **Figure 6**. We can see that the BMA performance is superior to the performance of individual model's output. The output of BMA has higher correlation with observed data than that of individual GCM's outputs. The BMA's result also has less standard deviation than individual model outputs and smaller RMSE than the individual GCMs' outputs. The other parts of **Figure 6** can be interpreted similarly. In the same way, other models can be evaluated with other variables mentioned in the above example and can help in model's selection in different areas.

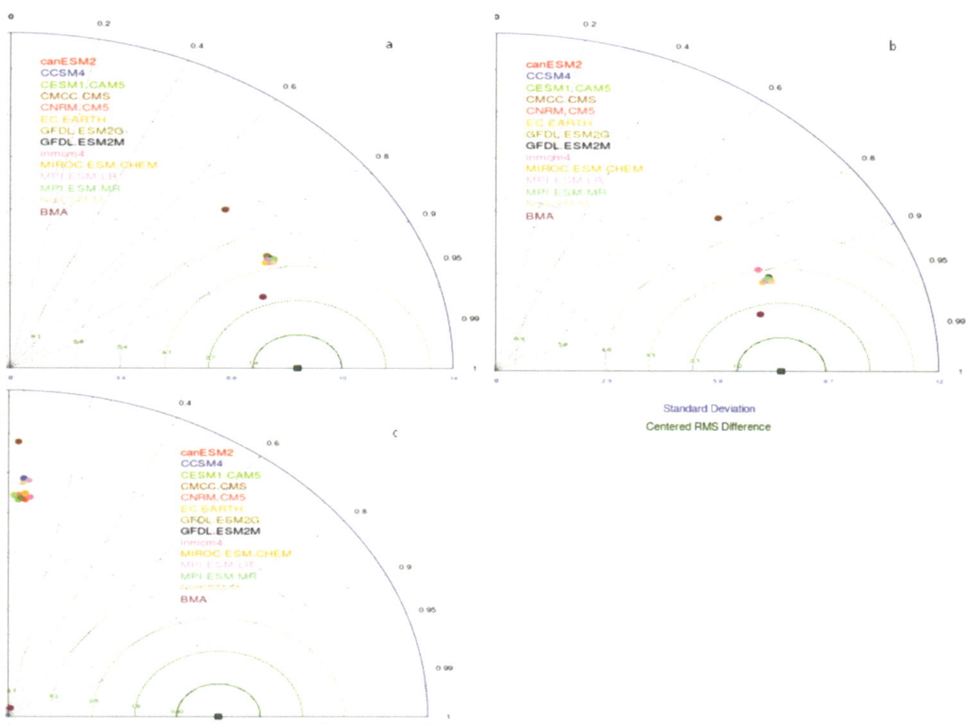

Figure 6. Taylor diagrams for evaluation of model and observational data for (a) maximum temperature, (b) minimum temperature, and (c) precipitation. Thirteen different GCMs' outputs and BMA's outputs are evaluated in this study for the Northern part of Pakistan.

4. Summary and conclusion

This study aims to present statistical methodology for evaluating process-based climate models. Different techniques have been presented for this purpose including statistical bias correction, meta analysis, model selection, ensemble projections, and Taylor diagram. The application of statistical bias correction bridged regional climate model's simulated and observational data. The performance of bias correction technique is better for temperature than precipitation; however, bias-corrected precipitation follows the observed precipitation's pattern nicely. Meta analysis can be used for different purposes like model selection, scenario analysis, etc. In this study, meta analysis is used for scenario analysis by considering four different scenarios two each from fourth assessment report (AR4) and fifth assessment report (AR5) of the IPCC. Meta analysis shows higher confidence in RCP projections and assigned higher weights on the basis of their precision. GCM's selection is of course important part in climate change assessment as there are many GCMs available. BMA is used for this purpose, and the results show that different variables have different ranks for different GCMs; however, they shared some GCMs in the list of best models. On the basis of selected GCM, ensemble projections were calculated using BMA technique. The results of GCMs and BMA's outputs were then evaluated by using Taylor diagram. Evaluation statistics used in Taylor diagram are root mean square error, correlation coefficient, and standard deviation of each data set. The evaluation confirms that ensemble projections are better than individual GCMs' outputs; nevertheless, we need to conduct this type of studies at different locations and then can make recommendations on the basis of their results.

Glossary

AEM	arithmetic ensemble mean
BMA	Bayesian model averaging
CC	coefficient of correlation
CORDEX	COordinated Regional climate Downscaling Experiment
FEM	fixed effect model
GCISC	Global Change Impact Studies Centre
GCM	global circulation/climate model
IPCC	Intergovernmental Panel on Climate Change
PDF	probability density function
PIP	posterior inclusion probability
PMD	Pakistan Meteorological Department
PPDF	predictive probability density function
RCM	regional climate model

REM	random effect model
RMSE	root mean square error
SD	standard deviation
UK	the United Kingdom

Author details

Firdos Khan[1,2]* and Jürgen Pilz[2]

*Address all correspondence to: fkyousafzai@gmail.com

1 Department of Mathematics and Statistics, International Islamic University, Islamabad, Pakistan

2 Department of Statistics, University of Klagenfurt, Klagenfurt, Austria

References

[1] Intergovernmental Panel on Climate Change. Available from: http://www.ipcc-data.org/guidelines/pages/gcm_guide.html [Accessed: 06-05-2018]

[2] Box GEP, Jenkins S. Time Series Analysis, Forecasting and Control. San Francisco: Olden Day; 1970

[3] Räty O, Räisänen J, Ylhäisi JS. Evaluation of delta change and bias correction methods for future daily precipitation: Intermodel cross-validation using ENSEMBLES simulations. Climate Dynamics. 2014;**42**:2287-2303. DOI: 10.1007/s00382-014-2130-8

[4] Rivington M, Matthews KB, Bellocchi G, Buchan K. Evaluating uncertainty introduced to process-based simulation model estimates by alternative sources of meteorological data. Agricultural Systems. 2006;**88**(2006):451-471. DOI: 10.1016/j.agsy.2005.07.004

[5] Lofan T, Dadson S, Buys G, Prudhomme C. Bias correction of daily precipitation simulated by a regional climate model: A comparison of methods. International Journal of Climatology. 2013;**33**:1367-1381. DOI: 10.1002/joc.3518

[6] Khan F, Pilz J, Amjad M, Wiberg DA. Climate variability and its impacts on water resources in the Upper Indus Basin under IPCC climate change scenarios. International Journal of Global Warming. 2015;**8**(1):46-69. DOI: 10.1504/IJGW.2015.071583

[7] Cannon AJ, Sobie SR, Murdock TQ. Bias correction of GCM precipitation by quantile mapping: How well do methods preserve changes in quantiles and extremes. Journal of Climate. 2015;**38**:6938-6959. DOI: 10.1175/JCLI-D-14-00754.1

[8] Cannon AJ. Multivariate quantile mapping bias correction: An N-dimensional probability density function transform for climate model simulations of multiple variables. Climate Dynamics. 2018;**50**(1-2):31-49. DOI: 10.1007/s00382-017-3580-6

[9] Borenstein M, Hedges LV, Higgins PT, Rothstein HR. A basic introduction to fixed-effect and random-effects models for meta-analysis. Research Synthesis Methods. 2010;**1**: 97-111. DOI: 10.1002/jrsm.12

[10] Chen DG, Peace KE. Applied Meta-Analysis with R. Boca Raton, Florida: A Chapman & Hall Book, CRC Press, Taylor & Francis Group; 2013

[11] Schwarzer G, Carpenter JR, Rücker G. Meta-analysis with R. New York: Springer International Publishing; 2015

[12] Viechtbauer W. Conducting meta-analyses in R with the meta for package. Journal of Statistical Software. 2010;**36**(3):1-48

[13] Beven K, Binley A. The future of distributed models—Model calibration and uncertainty prediction. Hydrological Processes. 1992;**6**(3):279-298. DOI: 0.1002/hyp.3360060305

[14] Blasone RS, Verugt JA, Madsen H, Rosbjerg D, Robinson BA, Zyvoloski GA. Generalized likelihood uncertainty estimation (GLUE) using adaptive Markov chain Monte Carlo sampling. Advances in Water Resources. 2008;**31**(4):630-648. DOI: 10.1016/j.advwatres.2007.12.003

[15] Freer J, Beven K, Ambroise B. Bayesian estimation of uncertainty in runoff prediction and the value of data: An application of the GLUE approach. Water Resources Research. 1996;**32**:2161-2173. DOI: 10.1029/95WR03723

[16] Stedinger JR, Vogel RM, Lee SU, Batchelder R. Appraisal of the generalized likelihood uncertainty estimation (GLUE) method. Water Resources Research. 2008;**44**:W00B06. DOI: 10.1029/2008WR006822

[17] Zheng Y, Keller AA. Uncertainty assessment in watershed-scale water quality modeling and management: 1. Framework and application of generalized likelihood uncertainty estimation (GLUE) approach. Water Resources Research. 2007;**43**:W08407. DOI: 10.1029/2006WR005345

[18] Uvsitalo L, Lehikoinen A, Helle I, Myrberg K. An overview of methods to evaluate uncertainty of deterministic models in decision support. Environmental Modelling & Software. 2015;**63**:24-31. DOI: 10.1016/j.envsoft.2014.09.017

[19] Lorenz EN. Atmospheric predictability experiments with large numerical model. Tellus. 1982;**34**:505-513

[20] Raftery AE, Gneiting T, Balabdaoui F, Polakowski M. Using Bayesian model averaging to calibrate forecast ensembles. Monthly Weather Review. 2005;**133**:1155-1174. DOI: 10.1175/MWR2906.1

[21] Min SK, Simonis D, Hense A. Probabilistic climate change predictions applying Bayesian model averaging. Philosophical Transactions of the Royal Society A. 2007;**365**:2103-2116. DOI: 10.1098/rsta.2007.2070

[22] Zeugner S. Bayesian model averaging with BMS. Technical Report. Available from: http://bms.zeugner.eu/tutorials/bms.pdf; 2011

[23] Hoeting JA, Madigan D, Raftery AE, Volinsky CT. Bayesian model averaging: A tutorial. Statistical Science. 1999;**14**(4):382-417

[24] Draper D. Assessment and propagation of model uncertainty. Journal of the Royal Statistical Society: Series B. 1995;**57**(1):45-97

[25] Huang Y. Comparison of general circulation model outputs and ensemble assessment of climate change using Bayesian approach. Global and Planetary Change. 2014;**122**:362-370

[26] Vrungt JA, Robinson BA. Treatment of uncertainty using ensemble methods: Comparison of sequential data assimilation and Bayesian model averaging. Water Resources Research. 2007;**43**:W01411. DOI: 10.1029/2005WR004838

[27] Duan Q, Ajami NK, Gao X, Sorooshian S. Multi-model ensemble hydrologic prediction using Bayesian model averaging. Advances in Water Resources. 2007;**30**(5):1371-1386

[28] Ajami NK, Duan Q, Gao X, Sorooshian S. Multimodel combination techniques for analysis of hydrological simulations: Application to distributed model intercomparison project results. Journal of Hydrometeorology. 2006;**7**:755-768

[29] Fernandez C, Ley E, Steel MFJ. Model uncertainty in cross-country growth regression. Journal of Applied Econometrics. 2001;**16**:563-576. DOI: 10.1002/jae.263

[30] Raftery AE, Zheng Y. Long-run performance of Bayesian model averaging. Journal of the American Statistical Association. 2003;**98**(464):931-938. DOI: 10.1198/016214503000000891

[31] Taylor KE. Summarizing multiple aspects of model performance in a single diagram. Journal of Geophysical Research. 2001;**106**:7183-7192. DOI: 10.1029/2000JD900719

Section 2

Impacts and Adaption

Chapter 5

The Impact of Climate Change on Water Availability and Recharge of Aquifers in the Jordan River Basin

Fayez Abdulla, Wim van Veen, Hani Abu Qdais,
Lia van Wesenbeeck and Ben Sonneveld

Additional information is available at the end of the chapter

http://dx.doi.org/10.5772/intechopen.80321

Abstract

Climate change can seriously affect the Middle East region by reduced and erratic rainfall. Formulating appropriate coping policies should account for local effects and changing flows interconnecting spatial units. We apply statistical downscaling techniques of coarse global circulation models to predict future rainfall patterns in the Yarmouk Basin, using a linear regression to extrapolate these results to the entire Jordan River Basin (JRB). Using a detailed water economy model for the JRB we predict rainfall patterns to evaluate the impact of climate change on agriculture and groundwater recharge. For the JRB, rainfall in 2050 will be around 10% lower than present precipitation, but with substantial spatial spreading. An overall reduction of net revenue from crop cultivation is estimated at 150 million USD, with major losses in Israel, Jordan, and the West Bank; Syrian revenues will slightly increase. The recharge of groundwater is affected negatively, and outflow to the Dead Sea is substantially lower, leading to further increases in salinization.

Keywords: Jordan basin, water economy model, climate change

1. Introduction

As is well-known, the current situation in the Jordan River Basin (JRB) is characterized by water scarcity and a history of water-related conflicts. The World Resources Institute (WRI) [1] classifies the JRB riparian countries among the most water-stressed countries in the world with a ratio of withdrawals to supply of more than 80%. Jordan is considered the poorest country in terms of water resources and most of its land is considered to be dry land. Except for the north-western highlands, 90% of the country receives less than 200 mm rainfall per year with an uneven distribution over regions and high fluctuation from year to year [2]. The population in the JRB region suffers from repeated water shortages which are more severe during the hot summers. Households devote considerable efforts to ensure their daily supply of water. In rural areas, the scarcity of water is among the main difficulties encountered by farmers and in urban areas, tap water is sometimes of bad quality, with frequent shortages. Households are forced to buy bottled or tank water at higher prices for their essential needs.

The severe water-related inequalities [3] are important drivers behind the regional conflicts and have created long-term political instability in the Middle East [4]. Rivalry has persisted over time as an imminent problem that has often been settled through force rather than peaceful cooperation [5]. Individual, uncoordinated actions by all riparian states have resulted in a dramatic change in water flows in the Jordan River Basin. According to United Nations Economic and Social Commission for Western Asia (UN-ESCWA) and Bundesanstalt für Geowissenschaften und Rohstoffe (BGR) [6], annual discharge of water into the Dead Sea under near-natural conditions would be approximately 1300 million m^3, but man-made interventions along the Jordan River and its tributaries have reduced this to 20–200 million m^3 at present. The massive reduction in water availability, particularly in the lower Jordan area, has fueled disputes between the riparian countries. Such disputes continue to obfuscate the relationships between Israel and Lebanon and between Israel and Syria [7]. Looking at the future, prospects for the region seem to be bleak. In the short run, the Syrian crisis with its large regional impact poses a serious threat to livelihoods and development. In the long run, climate change and a rapidly expanding population will continue to put more pressure on water resources.

Climate change is projected to have large impacts on weather patterns across the globe in the future. The Intergovernmental Panel on Climate Change (IPCC) in its 5th Assessment Report (AR5) noted that over the period 1800–2012, the average global temperature increased by 0.85°C. Moreover, Flato et al. [8] indicated that precipitation patterns deviate more frequently from long-term average trends in both volume and erraticism with adverse impacts on the society [9]. Realities of climate change are underpinned by Easterling et al. [10] who related climate change to increased drought incidence, flooding hazards, and reduced biodiversity. This clearly justifies calls for action to mitigate deteriorative effects of climate change. In this study, the focus is on the impact of climate change and water availability for households, effects on recharge to the aquifers, and on economic revenue in the Jordan River Basin, which as was argued above, is an area in the world where water scarcity is a major threat to economic growth and political stability.

Before we come up with practical and well-informed policy solutions, two challenges need to be addressed. First, the low resolution of global climate change models (100–500 km) ignores in-grid variability like complex topographical features and is, therefore, of limited use for impact studies [11]. Second, climate change effects cannot be restricted to changes in rainfall and temperature patterns alone but require a systemic response that accounts for spatial and temporal diversity of the natural resource base, interconnectedness of surface, and subsurface flows and influence on availability of water in volume and quality.

The first issue is addressed by using downscaled precipitation and climate parameters of Abdulla et al. [12] for meteorological stations in the Yarmouk Basin, covering parts of Syria and Jordan. The second challenge is met by using a water economy model of the JRB, which describes the natural and controlled flows in volume and quantity for Jordan River Basin as hydrological entity [13].

The remainder of this chapter is organized as follows. Section 2 gives a brief description of the structure, empirical basis, and calibration of the JRB water economy model. Section 3 presents the scenario formulation, including downscaling. Section 4 reports on the impact of climate change and Section 5 concludes and indicates pathways for further research.

2. The Jordan River Basin (JRB) model

Theoretically, the JRB model is a special case of the general class of welfare optimization models [13], where the innovative part is the inclusion of hydrology as central component of the production technology. Hence, control of flows (extraction of groundwater; use of water by humans; animals; and agriculture; transfer of water through canals; wastewater treatment and desalinization) conforms to basic principles of microeconomics with constraints that respect conservation. For economics, this implies that commodity balances hold; for hydrology, mass balances for pure water as well as for pollutants dissolved in water are maintained.

In its representation of the water economy of the JRB, the model distinguishes 48 districts, and 26 (two-weekly) time steps. Water can flow within and between five different layers. These comprise a surface layer on land for natural flows, a surface layer on land representing anthropogenic influences on water, a river layer, a root zone, and a layer representing the aquifer zones in the basin. Finally, next to clean water, the model can accommodate three types of pollutants: salinity, biological oxygen demand (BOD), and nitrate. The combination of place, time, layer, and quality defines a "cell" within the model, which acts as a source and destination of flows, representing the high level of interconnectedness of flows within the JRB.

2.1. Schematic representation of flows within the model

Figure 1 provides a schematic overview of flows within the JRB between the main layers, as well as flows entering the basin as whole (rainfall and lateral flows from outside the basin) or leaving the basin (evaporation, lateral flows leaving the basin). In the model, each of the layers and connections with other layers is modeled in detail, following hydrological laws as well as reflecting anthropogenic activity (pumping, irrigation, use by humans, livestock, industry and municipalities, sewage, waste water treatment and reuse, and water harvesting).

2.2. Water quality

As mentioned before, the JRB water economy model represents pollution of three specific types —salinity, nitrate, and BOD. Instead of using a set of attributes to represent water quality in each of these three dimensions at each point in time and space, the model represents these pollutants as flows, for which balances must hold, as for pure water. Hence, conceptually, pollutants are represented as flows with standard concentrations: 25 g/l for Cl, 2 g/l for NO_3, and 10 g/l for

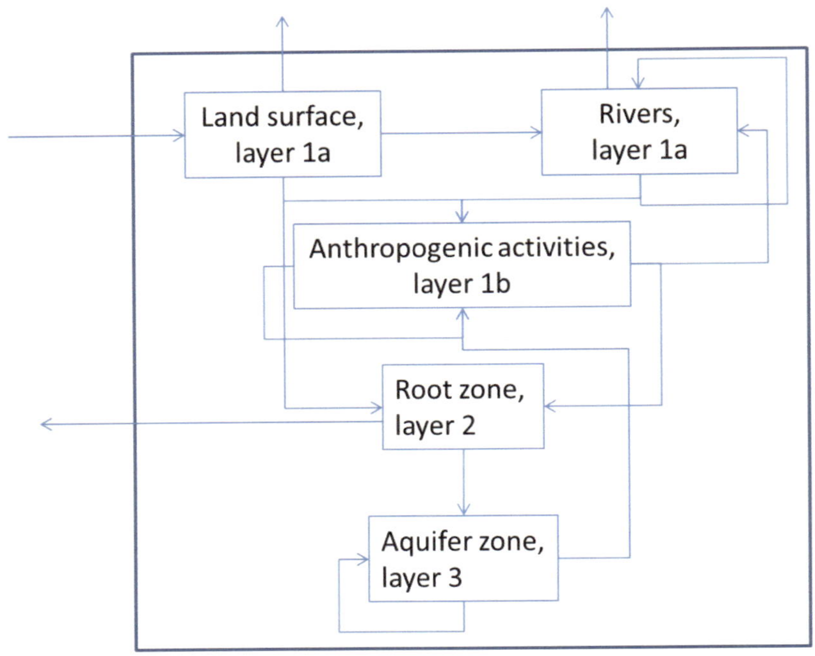

Figure 1. Schematic overview of flows within the JRB basin.

BOD. Observed concentrations are obtained by mixing the volumes of these flows with the volume of pure water. However, such a separation of flows would not do justice to the fact that pollutants flow with the water, and are not separable in reality. Hence, in the model application, flows are always combined. Where water quality changes, this is explicitly represented as a production process (water treatment) or a natural process (decreased quality through the uptake of pollutants from the soil). Consumers, livestock, and plants (irrigation) have to take the quality of water as it is at the location where it is offered, and this represents the channel through which quality changes enter into the system and affect yields, quality of produce and revenue.

2.3. Schematic representation of farmer behavior

Given the overriding importance of agriculture as economic activity in the JRB, the focus in assessing economic impact is on agriculture. In the JRB, crop cultivation is characterized by a dual system, where irrigation-based agriculture and rain-fed agriculture both exist. Hence, in principle, each farmer may have both types of land under cultivation. If profit maximization is taken as point of departure, a farmer in principle controls: (1) the amount of land under cultivation, (2) the share of the land that is irrigated, and (3) the crop(s) cultivated. In its simplest interpretation, profit maximization would be driven by the amount of water available for irrigation; the costs of inputs as well as irrigation water; and the output prices of different crops, while constraints would be defined by the response of different crops to the supply of water and other inputs.

The water response module of the model is richer than this simple representation, but also imposes some simplifying assumptions. To start with the latter, it is assumed that the crop composition observed in the base year 2010 is maintained under changing water availability. Furthermore, prices for inputs, water, and crops are assumed to remain constant, as are costs for irrigation. Richness of the module is achieved by acknowledging the following facts of agriculture in the JRB:

1. Whenever land is irrigated, irrigation applied is optimal for the crop under cultivation
2. Quality of land is not uniform over the JRB or even within districts
3. The highest-quality land is taken into production first; then, lesser qualities are used for cultivation
4. Water available for irrigation may contain (high levels of) salt, BOD, or nitrate
5. There is a large gap between yields on irrigated lands and yields on rain-fed areas
6. Yields on rain-fed areas mainly respond to changes in water availability
7. The area of rain-fed land does not respond to changes in water availability

This results in a decision tree for the farmer as depicted in **Figure 2**. For rain-fed lands, the farmer has no control over the response, as water availability in the root zone as well as water quality determine the yield on the given area. For irrigated lands, the farmer controls the area under irrigation (this block is marked purple, to indicate this is a decision variable). Yield response for the total area under irrigation is the result of expansion of irrigated areas to include land of lesser quality, where the optimal yield under irrigation is lower, leading to lower average yields for the total irrigated area. As for rain-fed agriculture, salinity has a negative impact on crop yields, and hence, also impacts on the yield response. Below, the chapter expands on the formal description of the water response module.

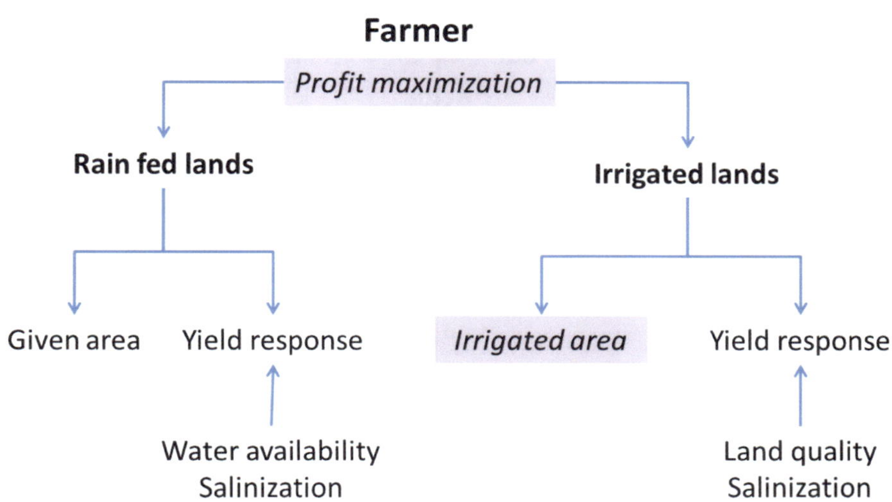

Figure 2. Farmer's response to changes in water quality and quantity.

2.4. Specification of the simulation model

Formally, the model can be represented by the following four equations:

$$d^i = \Delta^i x \tag{1}$$

$$e^j = H^j x \tag{2}$$

$$\tilde{x} = Rx \tag{3}$$

$$\sum_i d^i + \sum_j e^j + \tilde{x} = A\tilde{x} + \sum_i B^{0i} d^i + \sum_j B^j e^j + b \tag{4}$$

where b is the exogenous water availability (net rainfall, springs), d^i is water use of demand type i, e^j denotes input of project j, x represents water stock, $\tilde{x}$ is water volume available for natural outflow (including retention) and t represents time (two-weekly periods), while indices h are water quality classes, s are locations (district, river segment), ℓ are soil layers, i represents the type of water use and j the type of project. Finally, the following matrices of coefficients are defined in the model: A; the matrix of natural flow coefficients, B^{0i}; the matrix of return coefficients of demand type i, B^j; the matrix of return coefficients of project j; and Δ^i, a diagonal matrix with elements δ^i; H^j, a diagonal matrix with elements η^j; and R, a diagonal matrix with elements ρ.

Eq. (4) shows for each (h, s, t, ℓ) the balance between the destinations of the available water stock and the origins of this stock. Destinations are on the left-hand side, consisting of human water use, water inputs in projects and natural outflow. The latter includes retention, i.e., water of quality h remains at the same (s, ℓ) from t to $t + 1$. Origins of water availability are on the right-hand side, consisting of natural inflows, return flows from water use, return flows from projects, and exogenous availability from rainfall and springs.

Eqs. 1–3 specify each of the destinations as fixed fractions of the total available model stock x. Since retention is part of the natural outflows, the fractions logically have to add to one. Furthermore, the fractions have to reflect one important restriction, viz. that the user has to accept the quality of the available stock x. Therefore, the fractions do not depend on index h. More precisely, (1)–(3) can be written as:

$$d^i_{hst\ell} = \delta^i_{st\ell} \; x_{hst\ell} \tag{5}$$

$$e^j_{hst\ell} = \eta^j_{st\ell} \; x_{hst\ell} \tag{6}$$

$$\tilde{x}_{hst\ell} = \rho_{st\ell} \; x_{hst\ell} \tag{7}$$

2.5. Coverage of water flows

Although the equations look relatively simple, they cover a wealth of different flows especially due to the dimensionality of the system. We distinguish a hydrological component, anthropogenic activities, and actual human water use.

The hydrological component is represented by matrix A. Element $A_{hst\ell,h's't'\ell'}$ of matrix A denotes the share of $\tilde{x}_{h's't'\ell'}$ that flows in a natural way to (h, s, t, ℓ). Logically, the sum of these shares should not exceed one. In this way, the model can represent flows such as run-off of rainwater to the river system, percolation of rainwater to the root zone and further down to the groundwater zone, downstream flows from one river segment to the other or into Lake Tiberias, and retention of water at the same location and layer. Whenever relevant, changes in quality are also captured in this way, for instance taking up nitrate from the soil in run-off and percolation. If the sum of the outflow coefficients is less than one, it means that water disappears from the JRB. One may think in particular of evaporation or outflow to the Dead Sea. In the last period $(t = T)$, water disappears also via retention (becoming available next year).

In addition to these natural flows, the model has exogenous natural flows denoted as vector b. This parameter covers both net rainfall and water from springs. Net rainfall is measured after subtraction of immediate evaporation. It is important to be precise in this respect since immediate evaporation amounts to more than 10% of gross rainfall for the JRB as a whole. Spring water is essentially transferred from the groundwater zone to the surface zone of the district where it is used or to the river segment that it feeds. Hence, it has negative and positive values, summing to zero. For the first period $(t = 1)$, parameter b also covers the initial stock retained from the previous year.

Anthropogenic activities are represented by several projects j: groundwater pumping, rainwater harvesting, river outlets (dams), surface water transfer (canals, carriers), and irrigation. The input volumes of these projects are given by vectors e^j, each with dimensions $h,\ s,\ t, \ell$. The impact of the project is measured by elements $B^j_{hst\ell,h's't'\ell'}$, representing output (h, s, t, ℓ) per unit of input $\left(h^{'}, s^{'}, t^{'}, \ell^{'}\right)$. Projects that merely transfer water have allocation shares to a different layer or location (or period, in case of lags) for the same quality h. If water quality changes too, there are also allocation shares to quality classes $h^{'} \neq h$. Just as for matrix A, the sum of the output coefficients should not exceed one. Again, a sum strictly less than one means that water is disappearing from the JRB (evaporation, carrier to outside). Losses and leakages of projects are booked as output to other layers than intended, including evaporation.

Five types of water use are distinguished: household use, municipal use, industrial use, crop water use, and livestock use. Household water is further subdivided into tap water, truck water, roof water, and bottled water. Together, these destinations make up the set of water use types i. The public water distribution system delivers household tap water, municipal water, and industrial water. Eq. (1) specifies each type of water use as fraction of the available water stock. Therefore, d^i indicates gross water use, before purification.

When strictly following its definition above, matrix B^{0i} represents merely the return flows of used water, with element $B^{0i}_{h\ s\ t\ \ell,h's't'\ell'}$ denoting the return volume of (h, s, t, ℓ) per unit of water use $\left(h^{'}, s^{'}, t^{'}, \ell^{'}\right)$. However, since d^i indicates gross water use, matrix B^{0i} covers in fact a chain of activities: (a) water purification and distribution, (b) the actual return flow (production of waste water), (c) sewerage (collection of waste water), and (d) treatment of waste water.

To this, we may add even transport of raw and treated waste water. Hence, matrix B^{0i} is a composite of several matrices, although the number may differ across types of use i. Still, for B^{0i} the same condition as for A and B^j holds, viz. that the sum of its output coefficients should not exceed one. Distribution leakages and waste water dumping are booked as output to other layers than intended and where local quality changes result, this is also modeled as a pollution process.

2.6. Properties of the model

Here, we summarize the main properties of the model:

i. *No water creation:*

- for each (h, s, t, ℓ), the sum of the natural outflow coefficients does not exceed one
- for each (h, s, t, ℓ) and each project j, the sum of the project return coefficients does not exceed one
- for each (h, s, t, ℓ) and each type of use i, the sum of the demand return coefficients does not exceed one

ii. *Stock-driven flows:* all endogenous natural flows, all project inputs, and all volumes of water use are specified relative to the total available water stock

iii. *Blending property:* all users have to accept the quality of the available water stock; hence the fixed fractions in Eqs. (5)–(7) are independent of water quality class h

iv. *Exhaustive allocation of outflows*: the sum of the fixed fractions of all destinations, including those outside the JRB, is equal to one.

In model symbols the latter condition reads:

$$\sum\nolimits_i \delta^i_{st\ell} + \sum\nolimits_j \eta^j_{st\ell} + \rho_{st\ell} = 1 \tag{8}$$

Due to this condition, we can write Eq. (4) also fully in terms of total available stock x:

$$x = ARx + \sum\nolimits_i B^{0i}\Delta^i x + \sum\nolimits_j B^j H^j x + b \tag{9}$$

This formulation is used in the iterative calculation of the equilibrium stock levels that solve the model, starting from an initial value, say $x = x^0$. Due to the property of "no water creation," Eq. (9) represents a contraction mapping. Together with non-negativity of the parameters, this contraction property ensures convergence of the iterative calculations to a unique, non-negative fixed point x^*. The definition of a contraction mapping and the proof of this proposition are given in Chapter 6 of Keyzer[1] [14].

[1] In actual applications, this result will also hold if some of the elements of b are negative, provided that these negative values are overruled by other large positive values on the right-hand side of Eq. (6).

2.6.1. Crop revenue

Once the equilibrium stock x^* has been obtained, the corresponding values of d^i, e^j, and $\tilde{x}$ can also be calculated, including crop water use, taken up from the root zone, distinguished by location *s*, period *t*, and quality *h*. Based on the outcomes for crop water use, the impact on the net revenue of the farmer is calculated using additional exogenous information on the crop production structure in each district. This impact reflects both the direct impact of water on crop yields and the reaction of the farmer to the changes in water availability and water quality.

To this end, the simulation model includes a crop module that distinguishes K different crops, indexed *k*, and two land types, indexed *m*, viz. irrigated land (with or without protection) and rain-fed land. Water use and crop yields per hectare are different across these two land types. We introduce the following notation.

a_{kms} denotes harvested area of crop *k* on land type *m* in district *s*, per calendar year, in ha; $\bar{a}_{kms}$ is the reference area of crop *k* on land type *m* in district *s*, per calendar year, in ha; w_{kmhs} is the water use of quality *h* by crop *k* on land type *m* in district *s*, per calendar year, in m^3/ha; y_{kms} is the yield of crop *k* on land type *m* in district *s*, per calendar year, in kg/ha; $\bar{y}_{kms}$ is the reference yield of crop *k* on land type *m* in district *s*, per calendar year, in kg/ha; r_{kms} is net revenue of crop *k* on land type *m* in district *s*, per calendar year, in USD/ha; u_{kms} are the input costs (other than water) for crop *k* on land type in district *s*, per calendar year, in USD/ha; p^f_{ks} is the farm gate price of crop *k* in district *s*, in USD/kg; p^w_s is the water price paid by crop farmers in district *s*, in USD/m^3; and ρ_s is total net revenue from cropping in district *s*, per calendar year, in USD. The reference areas and yields refer to the observed 2010 levels.

Net revenue is calculated as follows:

$$r_{kms} = p^f_{ks} y_{kms} - u_{kms} - p^w_s \sum\nolimits_h w_{kmhs} \tag{10}$$

$$\rho_s = \sum\nolimits_{k,m} r_{kms} a_{kms} \tag{11}$$

The relation between actual harvested area a_{kms} and reference area $\bar{a}_{kms}$ and the relation between actual yield y_{kms} and reference yield $\bar{y}_{kms}$ are described by, respectively:

$$a_{kms} = \alpha_{kms}\ \bar{a}_{kms} \tag{12}$$

$$y_{kms} = \beta_{kms}\ \left(\gamma_{kms}/\bar{\gamma}_{kms}\right)\ \bar{y}_{kms} \tag{13}$$

In Eq. (12), α_{kms} represents the area adjustment due to the changes in water availability, expressed as factor relative to the reference area. For irrigated area, the factor depends positively on the water availability, irrespective of water quality. Depending on this volume, the factor is larger or smaller than one. For rain-fed area, the factor is by definition equal to one, hence no adjustment.

In Eq. (13), β_{kms} represents the yield adjustment factor due to changes in the availability of total water (irrespective of quality) while $\gamma_{kms}/\bar{\gamma}_{kms}$ measures the impact of changes in salinity. For irrigated area, factor β_{kms} is determined simultaneously with factor α_{kms} in (12), based on a subdivision of potentially irrigable land in land suitability classes. In case of extension of irrigated land, the factor may decline. For rain-fed area, factor β_{kms} is directly and positively related to water availability. Hence, for both land types, β_{kms} can be larger than or smaller than one.

Factor γ_{kms} measures the yield effect of salinity. It is calculated as a declining, piecewise linear function of the salinity level. By standardizing it to the 2010 reference value $\overline{\gamma}_{kms}$, the salinity impact in (13) is also a factor that can be larger than or smaller than one, just as factor β_{kms}.

In these equations, prices p^f_{ks} and p^w_s are exogenous, and so are reference areas $\overline{a}_{kms}$, reference yields $\overline{y}_{kms}$, and input costs u_{kms}, whereas factors α_{kms}, β_{kms}, and γ_{kms} depend on water volumes w_{kmhs}. Merbis and Sonneveld [15] describe the specification of these factors in more detail.

3. Scenario formulation

The following steps were taken to evaluate climate change impacts. First, we organized the daily precipitation data of the five climate change scenarios over the period 1980–2100 for 23 stations in the Yarmouk Basin in a excel spread sheet. Second, since variety of the modeled data are difficult to compare at daily level, we aggregated precipitation at yearly level and averaged the data of individual stations over cohorts of 20 consecutive years (2000–2020, 2021–2040, etc.) for each district. Third, we calculated relative changes over 40 (average of 2000–2020 minus 2040–2060) and 80 (average of 2000–2020 minus 2080–2100) years (**Table 1**). Fourth, using a linear regression that estimated a small positive gradient for the northern latitude [16], the predicted rainfall patterns for the Yarmouk river basin have been extrapolated to the entire JRB, constituting a database for the scenario analysis of daily rainfall at district level for the year 2050 (regression results are found in the Annex). Finally, we selected the CanESM2 model since, here, grid of the model overlaid best with the JRB area (in particular, better than CGCM3, see **Figure 3**); within CanESM2, RCP8.5 is selected as the most pessimistic scenario on future rainfall.

District	Altitude	North	Climann1 _40	Climann2 _40	Climann3 _40	Climann4 _40	Climann5 _40	Climann1 _80	Climann2 _80	Climann3 _80	Climann4 _80	Climann5 _80
CN21001	780	32,045	0,08	0,13	0,16	0,29	0,34	0,19	0,19	0,33	0,41	0,47
CN22021	-218	32,103	0,11	0,13	0,15	0,34	0,41	0,22	0,16	0,32	0,50	0,58
CN21010	715	32,299	0,08	0,20		0,43	0,46	0,21	0,25		0,51	0,56
CN21005	709	32,332	0,06	0,12	0,16	0,28	0,32	0,20	0,18	0,30	0,36	0,42
CN41004	1090	32,616	0,06	0,16	0,14	0,15	0,17	0,16	0,20	0,39	0,32	0,37
CN41007	618	32,791	0,06	0,14	0,12	0,09	0,09	0,16	0,18	0,31	0,22	0,28

Climann1_= CanESM2 GCM, RCP2.6, **Climann2_**= CanESM2 GCM, RCP4.5, **Climann3_**= CanESM2 GCM, RCP8.5, **Climann4_**= CGCM3 GCM, A1B, **Climann5_**= CGCM3 GCM, A2. **-40** = 40 years of difference, **_80** = 80 years of difference.

Table 1. Relative changes in annual precipitation by climate change scenario.

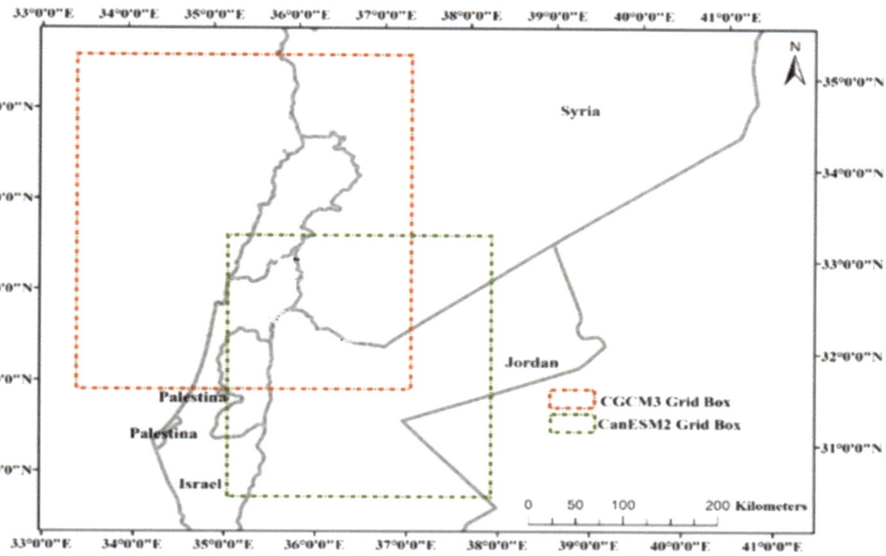

Figure 3. Grid overlay of CanESM2 and CGCM3 with JRB.

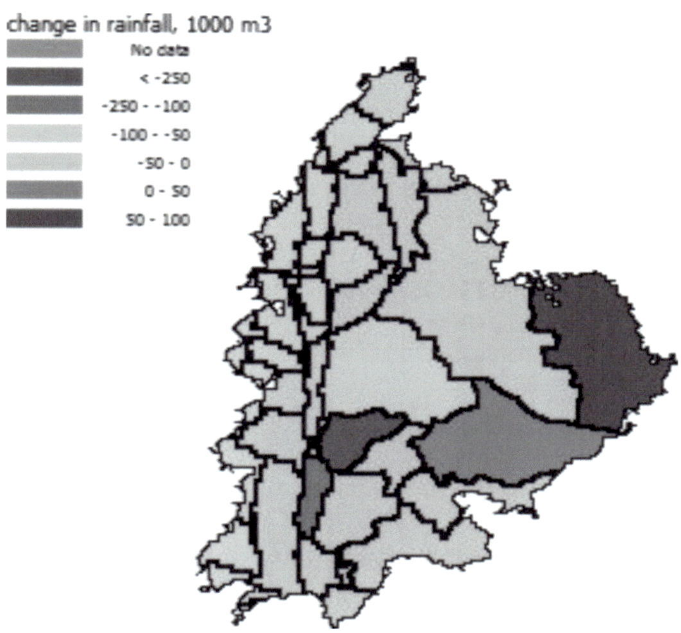

Figure 4. Rainfall in 2050: Predicted change relative to present conditions.

Following this procedure, we predict that for the whole of the JRB, rainfall in 2050 will be around 10% lower than present precipitation, but with substantial spatial spreading (see **Figure 4**), where rainfall is even predicted to increase in the eastern part of the JRB.

4. Impact of climate change

Application of the JRB model using these rainfall figures leads to the conclusion that the most important impact is an overall reduction of the net revenue from crop cultivation in the JRB as a whole of 150 million USD, with major losses in Israel, Jordan, and the West Bank. Syrian revenues would increase, reflecting the increase in rainfall in large parts of the country located in the JRB (see **Figure 5**).

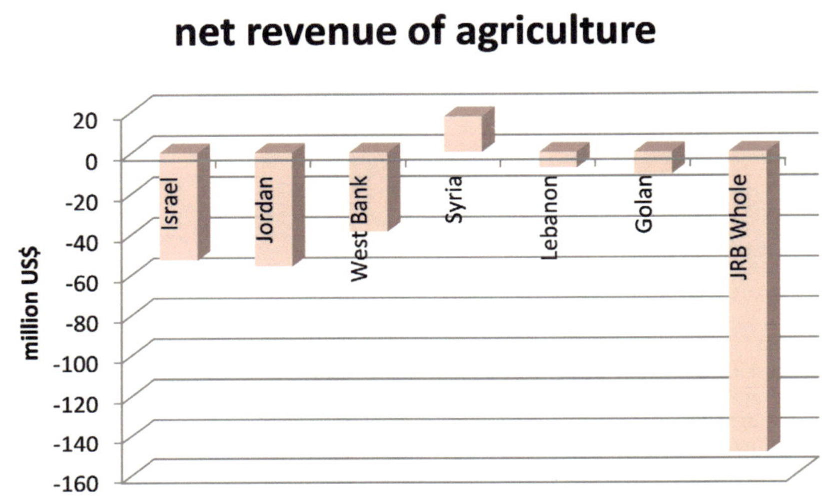

Figure 5. Impact of climate change on net revenue.

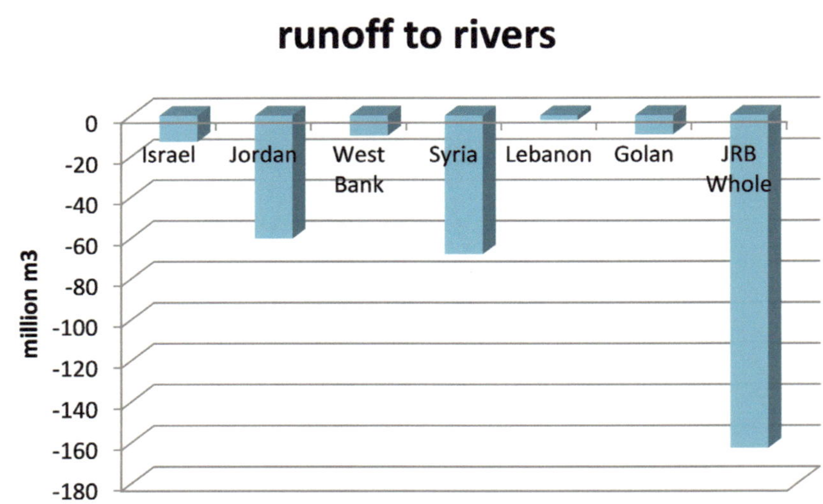

Figure 6. Impact of climate change on run-off.

However, the rainfall shock is not fully transmitted to agricultural activity: although the reduction in rainfall is a major shock, it is partly compensated by reduced evaporation (about half the shock), and lower river flows and hence lower extraction volumes. In addition, recharge of groundwater is affected very negatively, and the outflow to the Dead Sea is also substantially lower (20 million km^3). Despite the drop in recharge, salinity of groundwater resources increases only marginally (0.01% for the JRB as a whole, with a "peak" of 0.02% for Israel). Decreasing availability of water implies a decrease in the amount of untreated waste water, as household

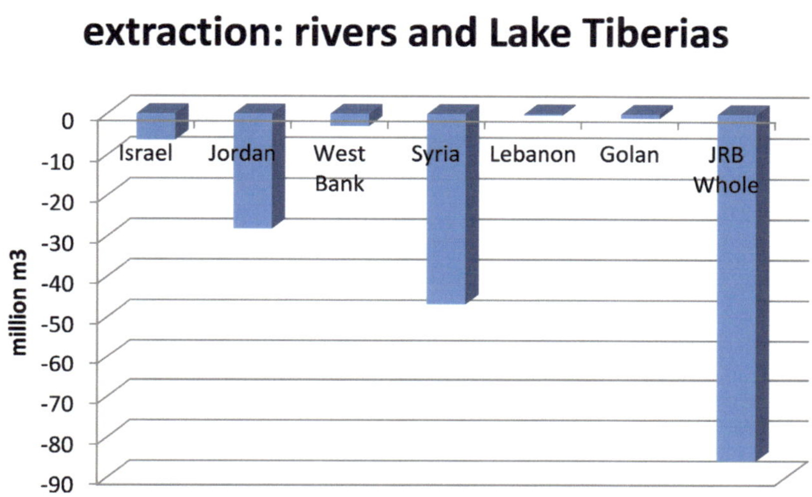

Figure 7. Impact of climate change on extraction.

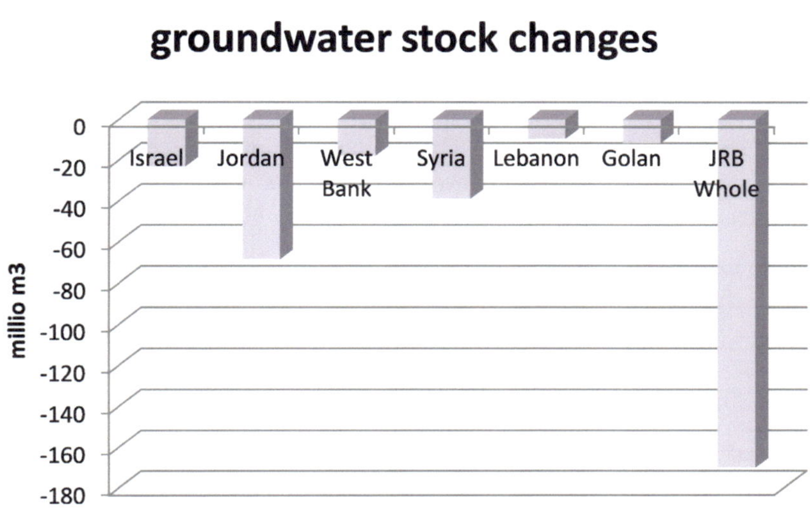

Figure 8. Impact of climate change on groundwater stock changes.

demand falls and less waste water is produced. This in turn implies that irrigation water contains relatively less untreated waste water and that leakage of contaminated water to the root zone also decreases, leading to a reduction in BOD in water used for irrigation of 5.5% on average for the JRB. **Figures 6–8** illustrate the impact on run-off to the rivers, extraction from the rivers and Lake Tiberias and groundwater stock changes relative to the present baseline. Particularly, the latter result is cause for concern about the future as it is clear that unchanged extraction policies would lead to unsustainable extraction from groundwater resources in the coming decade.

5. Conclusions and way forward

Climate change is a major concern for policy makers in the Middle East who aim to protect their constituency from adverse effects on water availability for food security and the environment. Yet, to provide a proper foundation for informed decisions on adoption or mitigation of climate change effects, two issues have to be resolved. First, results of global circulation models should be downscaled from their coarse (50–100 km) grids to a meaningful spatial resolution. Second, effects of changes in rainfall should reflect the hydrological complexity of natural and controlled surface and subsurface flows that jointly are responsible for water supply.

In this study, we addressed these two issues for the JRB by a statistical downscaling of the climate change scenarios for precipitation from global circulation models in conjunction with the application of a water economy model that describes the water flows in detail and, additionally, reports on the impact of climate change on water availability (run-off and groundwater recharge) and agricultural productivity in JRB.

The results reveal that there is a significant reduction in the surface run-off of an amount greater than 160 MCM (28%) in the JRB that affects the extraction volumes out of the rivers. Most reduction in run-off to rivers is found in Syria (42%) and Jordan (40%). There is also a significant reduction in groundwater recharge of an amount of about 180 MCM (11%), which seriously threatens the already overexploited groundwater stock. Accordingly, the agricultural productivity reduces by 160 million USD, with largest effects on Jordan (23%), the West Bank (16%), and Israel (12%). The impact of climate change on water quality is minor; quality even improves due to lower waste water volumes that blend with fresh water resources.

The results obtained in this study could be used as a reference for regional water resources management in the JRB. The results also provide detailed information on spatially explicit effects that allow local policy makers to take matters in their own hand. As such, the model outcomes can also underpin the stakeholder discussions on distribution of water resources to support negotiations on water transfers, within and between basins.

Further validation exercises should strengthen the reliability of results obtained by the global climatic models (GCMs). Incorporation of local knowledge on water management that transcends

http://dx.doi.org/10.5772/intechopen.80321

academic disciplines with practical solutions and on the ground reality further strengthens the water economy model's representation and its utility as decision support tool.

A. Appendix

See **Figures a1–a10**.

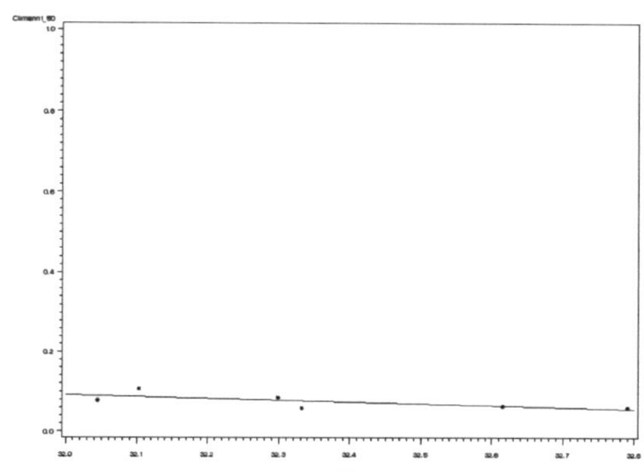

Figure a1. Relative future changes in precipiation against South-North gradient. Scenario: CanESM2 GCM, RCP2.6, time lapse 40 years.

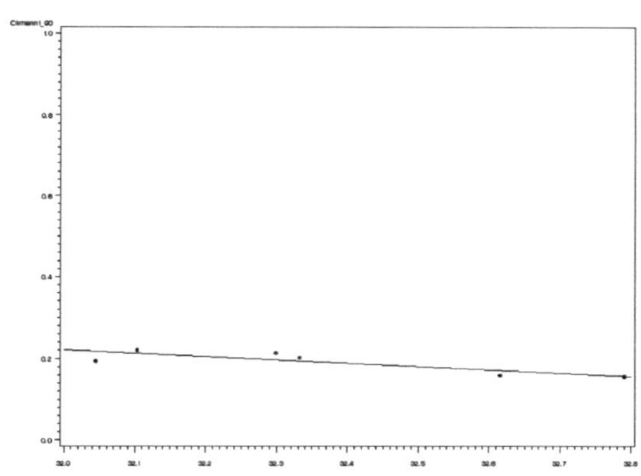

Figure a2. Relative future changes in precipiation against South-North gradient. Scenario: CanESM2 GCM, RCP2.6, time lapse 80 years.

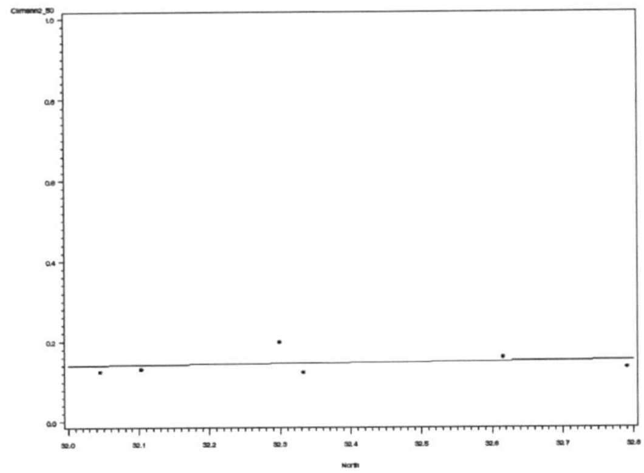

Figure a3. Relative future changes in precipiation against South-North gradient. Scenario: CanESM2 GCM, RCP4.5, time lapse 40 years.

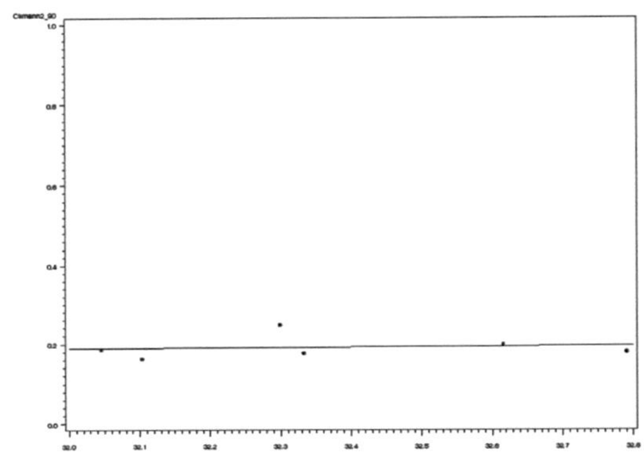

Figure a4. Relative future changes in precipiation against South-North gradient. Scenario: CanESM2 GCM, RCP4.5, time lapse 80 years.

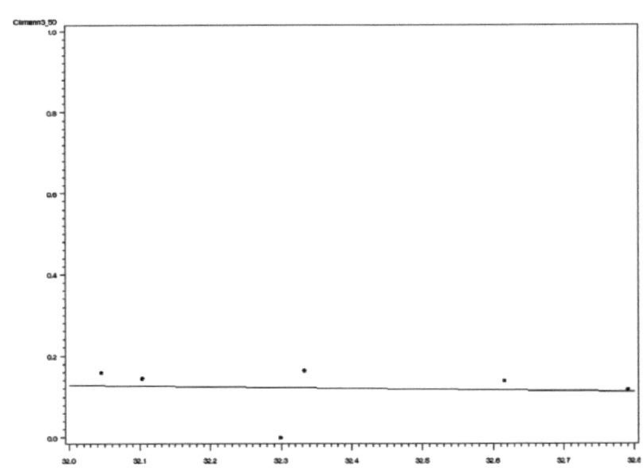

Figure a5. Relative future changes in precipiation against South-North gradient. Scenario: CanESM2 GCM, RCP8.5, time lapse 40 years.

http://dx.doi.org/10.5772/intechopen.80321

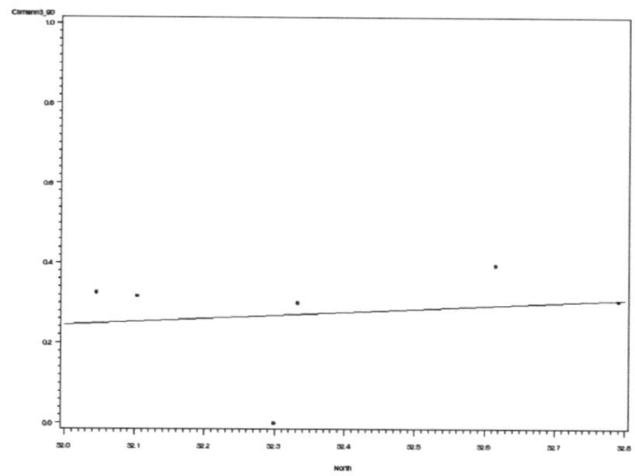

Figure a6. Relative future changes in precipiation against South-North gradient. Scenario: CanESM2 GCM, RCP8.5, time lapse 80 years.

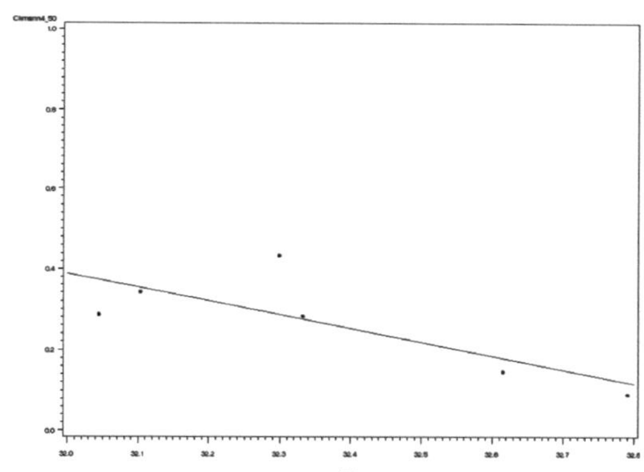

Figure a7. Relative future changes in precipiation against South-North gradient. Scenario: CGCM3 GCM, A1B, time lapse 40 years.

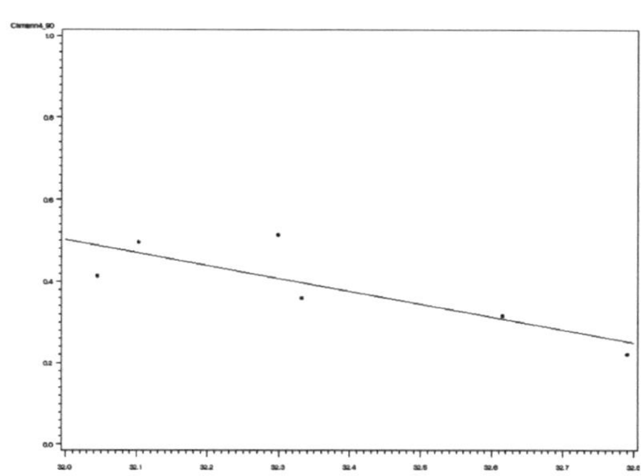

Figure a8. Relative future changes in precipiation against South-North gradient. Scenario: CGCM3 GCM, A1B, time lapse 80 years.

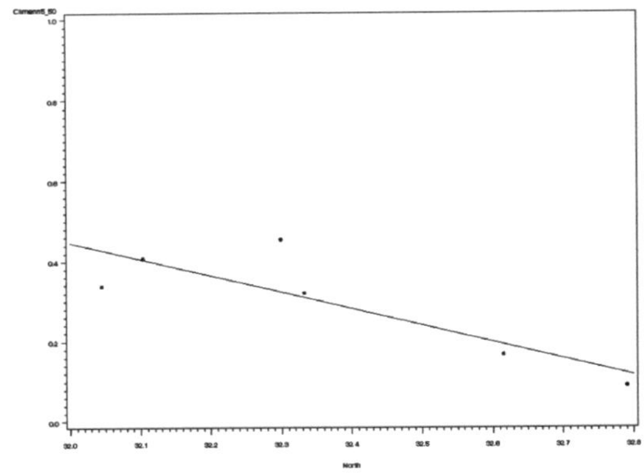

Figure a9. Relative future changes in precipiation against South-North gradient. Scenario: CGCM3 GCM, A2, time lapse 40 years.

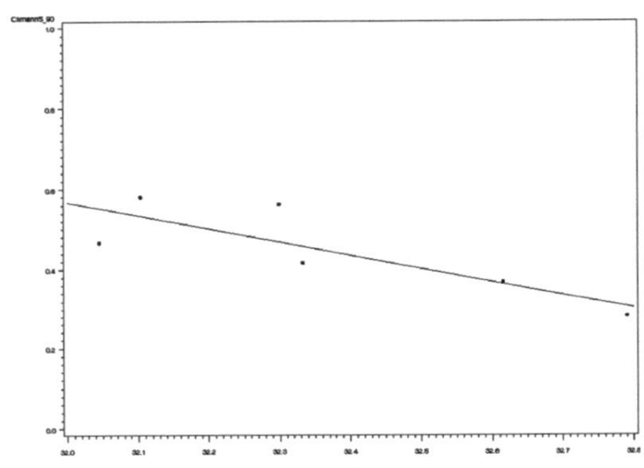

Figure a10. Relative future changes in precipiation against South-North gradient. Scenario: CGCM3 GCM, A2, time lapse 80 years.

Author details

Fayez Abdulla[1], Wim van Veen[2], Hani Abu Qdais[1], Lia van Wesenbeeck[2*] and Ben Sonneveld[2]

*Address all correspondence to: c.f.a.van.wesenbeeck@vu.nl

1 Civil Engineering Department, Jordan University of Science and Technology, Irbid, Jordan

2 Amsterdam Centre for World Food Studies, VU University, Amsterdam, The Netherlands

References

[1] WRI. World's 36 Most Water-Stressed Countries. Washington: World Resource Institute; 2013 http://www.wri.org/blog/2013/12/world%E2%80%99s-36-most-water-stressed-countries

[2] Abdulla FA, Eshtawi T, Assaf H. Assessment of the impact of potential climate change on the water balance of semi-arid watershed. Water Resources Management. 2009;**23**:2051-2068

[3] Selby J. Cooperation, domination and colonisation: The 'region a'-Palestinian joint water committee. Water Alternatives. 2013;**6**:1-24

[4] Wolf AT. Hydropolitics along the Jordan River; Scarce Water and Its Impact on the Arab—'Region a' Conflict, Vol. 99. United Nations University Press; 1995

[5] Ashok S, Jägerskog A. Emerging Security Threats in the Middle East: The Impact of Climate Change and Globalization. Rowman & Littlefield; 2016

[6] UN-ESCWA, BGR. Inventory of Shared Water Resources in Western Asia, Chapter 6, Jordan River Basin. Beirut: United Nations Economic and Social Commission for Western Asia; Bundesanstalt für Geowissenschaften und Rohstoffe; 2013

[7] Zeitoun M, Talhami M, Eid-Sabbagh K. The influence of narratives on negotiations and resolution of the upper Jordan River conflict. International Negotiation. 2013;**18**:293-322

[8] Flato G, Marotzke J, Abiodun B, Braconnot P, Chou SC, Collins W, et al. Evaluation of climate models. In: Stocker TF, Qin D, Plattner G-K, Tignor M, Allen SK, Doschung J, Nauels A, Xia Y, Bex V, Midgley PM, editors. Climate Change 2013: The Physical Science Basis. Contribution of Working Group I to the Fifth Assessment Report of the Intergovernmental Panel on Climate Change. Cambridge University Press; 2013. pp. 741-882. DOI: 10.1017/CBO9781107415324.020

[9] IPCC. Managing the risks of extreme events and disasters to advance climate change adaptation. In: Field CB, Barros V, Stocker TF, Qin D, Dokken DJ, Ebi KL, Mastrandrea MD, Mach KJ, Plattner G-K, Allen SK, Tignor M, Midgley PM, editors. A Special Report of Working Groups I and II of the Intergovernmental Panel on Climate Change. Cambridge, UK, and New York, NY, USA: Cambridge University Press; 2012. 582 pp

[10] Easterling DR, Evans JL, Groisman PY, Karl TR, Kunkel KE, Ambenje P. Observed variability and trends in extreme climate events: A brief review. Bulletin of the American Meteorological Society. 2000;**81**:417-425

[11] Gu H, Wang G, Yu Z, Mei R. Assessing future climate changes and extreme indicators in east and south Asia using the RegCM4 regional climate model. Clim. Change. 2012;**114**(2): 301-317. DOI: 10.1007/s10584-012-0411-y

[12] Abdulla F., H. Abu Qdais, A. Al-Shurafat, et al. (2016) Future changes in precipitation and maximum and minimum temperature using the statistical downscaling model (SDSM) in the trans-boundary Yarmouk River Basin

[13] van Veen W, van Wesenbeeck L, Merbis M, Sonneveld B. Towards concerted sharing: Development of a regional water economy model in the Jordan River Basin. Final Report. Centre for World Food Studies of the VU University Amsterdam; 2017

[14] Keyzer MA. Optimal calibration and control in a regional water economy model for the Jordan River basin. Working Paper under the project. Towards concerted sharing: Development of

a regional water economy model in the Jordan River Basin. Amsterdam, Netherlands: Centre for World Food Studies; 2015

[15] Merbis MD, Sonneveld B. Construction of water response functions in the Jordan River basin. Working paper under the project. Towards concerted sharing: Development of a regional water economy model in the Jordan River Basin. Amsterdam, Netherlands: Centre for World Food Studies; 2016

[16] McKinney W. Data structures for statistical computing in Python. In: Proceedings of the 9th Python in Science Conference. 2010. pp. 51-56

Chapter 6

Super Typhoon Bopha and the Mayo River Debris-Flow Disaster, Mindanao, Philippines, December 2012

Kelvin S. Rodolfo, A. Mahar F. Lagmay,
Rodrigo C. Eco, Tatum Miko L. Herrero,
Jerico E. Mendoza, Likha G. Minimo, Joy T. Santiago,
Jenalyn Alconis-Ayco, Eric C. Colmenares,
Jasmine J. Sabado and Ryanne Wayne Serrado

Additional information is available at the end of the chapter

http://dx.doi.org/10.5772/intechopen.81669

Abstract

Category 5 (C5) Super Typhoon Bopha, the world's worst storm of 2012, formed abnormally close to the West Pacific Equator, and Bopha's Mindanao landfall has the record equatorial proximity for C5 storms. Bopha generated a debris flow that buried 500 ha of New Bataan municipality and killed 566 people. New Bataan, established in 1968, had never experienced super typhoons and debris flows. We describe the respective histories of New Bataan and Super Typhoon Bopha; debris flows; and how population growth and unwise settlement practices contribute to Philippine "natural" disasters. The historical record of Mindanao tropical cyclones yields clues regarding how climate change may be exacerbating near-equatorial vulnerability to typhoons. Existing models of future typhoon behavior do not apply well to Mindanao because they evaluate only the tropical cyclones that occur during the main June–October typhoon season, and most Mindanao tropical cyclones occur in the off season. The models also ignore tropical depressions, the most frequent—and commonly lethal—Mindanao cyclones. Including these in annual tallies of Mindanao cyclones up to early 2018 reveals a pronounced and accelerating increase since 1990. Mindanao is susceptible to other natural hazards, including other consequences of climate change and volcanic activity.

Keywords: Super Typhoon Bopha, Andap disaster, Mayo River, debris flows, climate change, ENSO, typhoon frequency

IntechOpen

1. Introduction

On 4 November, 2012, Super Typhoon Bopha generated a massive debris flow that devastated *barangay* (village) Andap in the Mindanao municipality of New Bataan and killed hundreds of people. In early 2013, we were designated as a field disaster-analysis team by Project NOAH (Nationwide Operational Assessment of Hazards), the disaster-assessment program of the University of the Philippines in Diliman, Quezon City.

Prior to our field work, we gathered high-resolution optical satellite imagery for mapping out the extent of the debris flow deposits and commissioned a Light Detection and Ranging (LiDAR) survey to generate detailed topographic maps of the area. In the field, we analyzed and plotted the new deposits on our new maps. They were clearly left by a debris flow, and we determined its velocity when it hit Andap from scarring on impacted trees. Old deposits were left in the area by debris flows that occurred long before New Bataan was established. Eyewitnesses recounted the Bopha event for us in detail, and long-time residents informed us that similar events had never happened before. We analyzed and reconstructed the event from all these gathered data.

An initial report we published in 2016 [1] described the Super Typhoon, the Mayo River debris flow, and the detailed geologic reasons for it. We also discussed how and reviewed how population growth and inadequate geological analysis of settlement sites contribute to Philippine "natural" disasters. Our report discussed how climate change may be bringing more frequent major typhoons and debris flows they trigger to Mindanao and to other vulnerable subequatorial areas. We did so by examining the sparse record of tropical cyclones that made landfall on Mindanao since 1945, associated records of the Pacific El Niño-Southern Oscillation (ENSO), and all western North Pacific tropical cyclones from 1945 to 2015.

Here, we update that evaluation with additional data from 2016 through February 2018. A positive outgrowth of this research is Project NOAH's new program that has identified more than a thousand Philippine alluvial fans and associated communities that might experience debris flows. This program already helped to mitigate debris flows on Luzon and Mindoro islands. We conclude by exploring possible protective measures for climate-related hazards that threaten Mindanao and other subequatorial areas.

2. Prolog to disaster: geomorphologic setting and history of New Bataan

Southeastern Mindanao is a rugged coastal range (**Figure 1**). About 35 km west of the coast, and 3 km upstream of Andap, the Mayo River drains a rugged, 36.5 km^2 watershed on the western slopes of the coastal range. In the Mayo watershed, many slopes are steeper than 35° and total relief is about 2320 m. Flowing northward, the Mayo River debouches through a narrow gorge to join the Kalyawan River, which flows northward along the Compostela Valley, as do other Agusan River southern tributaries.

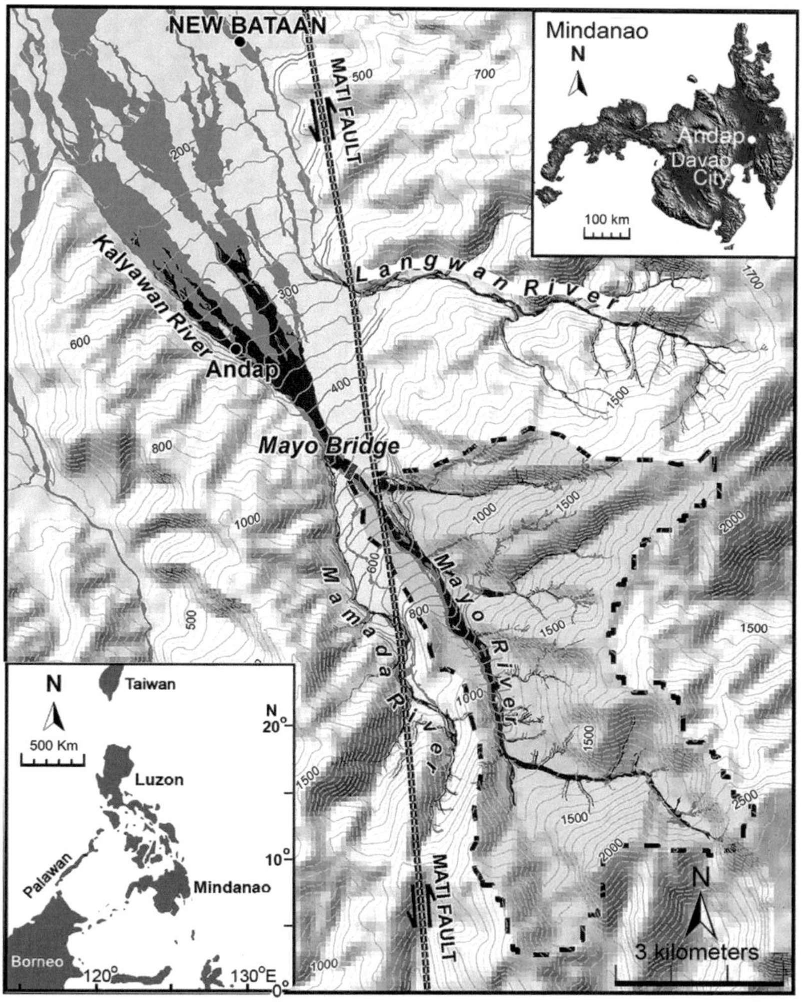

Figure 1. Physical setting of the Andap disaster. Gray area enclosed by dashes is the Mayo River watershed. All steep slopes are contoured at 50-m intervals. Below 700-m elevations, the contour interval is 20 m to better define the gentler valley surfaces. New deposits of "true" debris flows are shown in solid black; associated hyperconcentrated-flow deposits are shaded in gray. Note that the topographic contour lines from the Mayo Bridge to Andap are convex northward, defining the surface of an alluvial fan just upstream from Andap. The trace of the Mati Fault is only generalized; it has numerous associated fractures in a broad zone along its length.

A site 8 km below the Mayo-Kalyawan junction in the eastern Compostela Valley called "Cabinuangan" because of its many huge Binuang (*Octomeles sumatrana*) trees began to be logged in the early 1950s [2]. As the loggers rapidly expanded their road networks, immigrant farmers from Luzon and the Visayan Islands followed closely behind, planting the cleared land mainly to coconuts, but also to rice, corn, bananas, coffee, cacao, abaca, and bamboo.

The Philippine government divided the public lands of Compostela Valley into formal municipal areas beginning in 1966. One covering 55,315 ha in Cabinuangan was named New

Bataan in 1968 because Luz Banzon-Magsaysay, a native of the Luzon province of Bataan and President Magsaysay's widow, had espoused its establishment. New Bataan was subdivided into 16 *barangays* (villages) comprising farm lots. A 154-ha area at the center of New Bataan was designated the town site and given the *barangay* name of "Cabinuangan." In 1970, 2 years after its founding, the population of New Bataan was 19,978 [3]; by 1 May, 2010, it had increased 238% to 47,470, including 10,390 in Cabinuangan and 7550 in Andap [4].

The town planners made a nice design for Cabinuangan, its streets fanning out geometrically from its central core of government and social buildings (**Figure 2A**). Unfortunately, the planners knew little about natural hazards. Even government authorities did not know that the Kalyawan River had been a conduit for ancient debris flows; as late as 2012, the official hazard map of New Bataan [5] evaluated only landslide and flood risks. This lack of geomorphologic knowledge was fatal during Bopha (**Figure 2B**).

Barangay Andap was established at the head of Compostela Valley on high ground 3 km upstream of Cabinuangan. That site was not recognized as an alluvial fan, a landform built up by successive debris flows. Our field work documented that the fan was built up by characteristically reverse-graded, matrix-supported debris-flow deposits of unknown but ancient age (**Figure 3**).

2.1. Debris flows

Among the world's most destructive natural phenomena, debris flows are fast-moving slurries of water and rock fragments, soil, and mud [6–9]. Many debris flows (**Table 1**) [10] are associated with volcanoes [11, 12]; many others are not, including the Mayo River event. All that is required to generate a debris flow is an abundance of loose rock debris and soil and a sudden large influx of water. They can be triggered by sudden downpours such as commonly delivered by tropical cyclones, by reservoir collapses [13], or by landslides dislodged by earthquakes into streams.

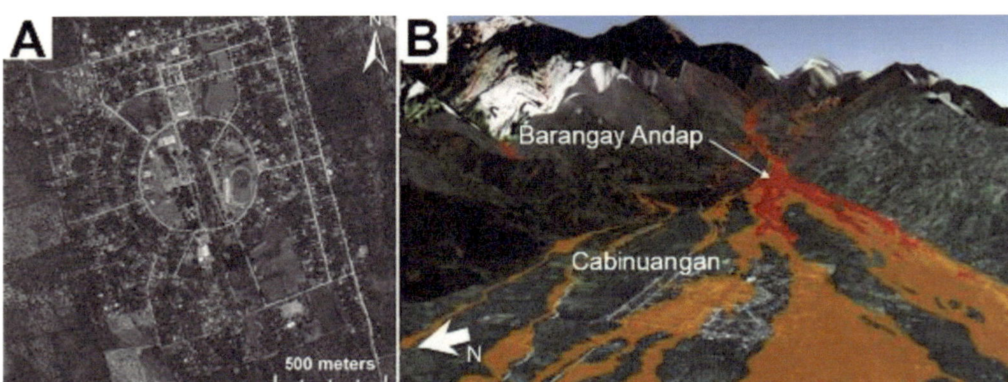

Figure 2. New Bataan. A= Andap, Google image of Cabinuangan (the central district of New Bataan) before the debris flow. B= Southward facing three-dimensional terrain diagram of Andap and Cabinuanga after the Mayo River disaster. Red areas are boulder-rich "true debris flow; orange areas are deposits of more dilute " hyperconcentrated" flows.

Figure 3. Debris-flow deposits in the New Bataan area. (A) Boulder in ancient reverse-graded debris-flow deposit. Well-established trees indicate an age of some decades prior to the settlement of the town. (B) Old debris-flow deposits underlying New Bataan—Andap high-way. Boulders and cobbles are separated from each other by a matrix of finer-grained sediment, as they were while still flowing. For scale, the concrete is 15-cm thick. The coarse sediments atop the highway are new debris-flow deposits from Typhoon Bopha. (C) Boulder-rich deposits of debris flows that destroyed much of the barangay, at the site of the destroyed Mayo River bridge.

The lethality and capacity for damage of a debris flow is not determined by its size alone. If its path is sparsely populated, such as at Mount St. Helens, or if the people in harm's way are familiar with the hazard, such as at Pinatubo Volcano, even large debris flows may not inflict casualties.

Rain on mountain slopes that falls strongly and lasts long enough will dislodge soil and loose rock into landslides. These may coalesce into debris flows, which are slurries of sediment and water that look and behave like concrete pouring out of a delivery truck. By weight, the water rarely exceeds 25%; only 10% may be enough to provide mobility. Gravel and boulders constitute more than half of the solids, and sand typically makes up about 40%. Silt and clay normally constitute less than 10% and remain suspended in the water [21, 22]. Students of debris flows frequently say "In stream floods, the water carries the sediment; in debris flows, the sediment carries the water."

While a debris flow is contained in a mountain channel, it carries large boulders with remarkable ease. In part, this is because of the high buoyancy of the dense slurry. Additionally, boulders in the flow repeatedly bounce away from the channel floor and sides up into the "central plug" of the flow near the surface, where friction with the channel is minimal and the

Location	Date	Trigger	Volume, 10^6 m^3	Deaths
Rios Barrancas and Colorado, Argentina [14]	1914	Failure of ancient landslide dam	2000 estimated	?
Bucao River, Pinatubo Volcano, Philippines [13]	10 July, 2002	Caldera lake breach	<<160	0
Bucao River, Pinatubo Volcano, Philippines [15]	5–6 Oct, 1993	Typhoon Flo (Kadiang) rains	110	0
Kolka Glacier, North Ossetia, Russia [16]	2002	Large glacial detachment	~100	125
Nevados Huarascan, Peru [17]	1970	Pyroclastic flows melted snow and ice	100 (flow volume)	18,000
Nevado del Ruiz, Colombia [14]	13 Nov, 1985	Pyroclastic flows melted snow and ice	40	23,000
Mayo River, Mindanao, Philippines [1]	4 Dec, 2012	Typhoon Bopha (Pablo) rainfall	25–30	566
Cordillera de la Costa, Vargas, Valenzuela [18]	Dec 1999	Heavy rain	19	30,000
Mayon Volcano, Philippines [19]	30 Nov, 2006	Typhoon Durian (Reming) rains	19	1226
Pine Creek—Muddy River, Mount St. Helens, Washington, USA [20]	18 May, 1980	Pyroclastic surge melted snow & ice	14	0

Table 1. The global record of the 10 largest debris flows, ranked by decreasing volume. Modified and updated from [10].

flow is fastest, enabling them to migrate quickly to the front of the flow. There, they become part of a moving dam of boulders, logs, and tree debris being pushed along by the flowing mass contained behind it.

The moving frontal dam ponds the main flow body, which is richer in sand, silt, and clay and progressively becomes more dilute toward the rear, undergoing transitions into what are called *hyperconcentrated* flows, somewhat confusingly because they carry much more sediment than do normal streams. In hyperconcentrated flows, sand, silt, and clay typically comprise up to 75% by weight. Such flows look like normal, turbid flood waters, but their velocities are much greater, typically 2–3 m/s [23]. They are too dilute to transport boulders and can transport gravel only by pushing and rolling it on the channel floor. To the rear, hyperconcentrated flows are succeeded by even more dilute, turbid flood water. In the literature, somewhat confusingly, "debris flow" sometimes refers to only a true debris-flow phase. Sometimes, however, the term means an entire hydrologic event consisting of debris-flow, hyperconcentrated, and normal stream-flow phases, as we do here in reference to the Mayo River debris flow.

When a debris flow emerges from the mountains, it spreads out, and the increased basal friction slows it down. Some of its sediment load drops out and adds volume to an *alluvial fan*, a cone-shaped feature that topographic maps show as contour lines that are convex in the downstream direction, as seen in **Figure 1**. Even after the debris flow spreads out, large boulders (**Figure 3**) continue to be transported by combined flotation, push, drag, and rolling. The hyperconcentrated and normal-flood phases may extend many kilometers beyond the alluvial fan. Debris flows vary in volume by many orders of magnitude (**Table 1**), the most frequent ones being only a 1000–100,000 m^3 and the largest more than a 100,000,000 m^3 [10].

An important distinguishing characteristic of true debris-flow deposits is "reverse grading": boulders tend to be smaller at the base and increase in size upwards. Large boulders commonly jut out at the top of a deposit, as observed at New Bataan (**Figure 3A**). In addition to the buoyancy they experience from the dense slurry, the best mechanism advanced to explain reverse grading is *kinetic sieving* [7, 11, 24–26]. While flowing, shear at the base of a debris flow continuously causes temporary void spaces of different sizes to open, and particles of equivalent sizes migrate into them. Smaller voids form and are filled by smaller solid particles more frequently, and so larger boulders migrate up toward the top of the flow. Another characteristic of debris-flow deposits that distinguish them from the deposits left by normal streams, in which particles grade upward from coarse to fine, is "matrix support" (**Figure 3B**). A mixture of the finer sediment that constituted the bulk of the flow separates the larger rock fragments from each other. A useful guide for distinguishing the effects of debris flows from those of floods was published by Pierson [27].

3. Super Typhoon Bopha

On 23 November, 2012, a large area of convection began forming at 0.6°N latitude, 158°E longitude [28] (**Figure 4A**). Two days later, while still unusually close to the equator at 03.6°N, 157°E, it was categorized as a tropical depression. It was upgraded to Tropical Storm Bopha three days later on 26 November, while at 04.4°N, 155.8°E, a latitude where the Coriolis effect was too weak to quickly cause it to rotate. Only four days later, on 30 November, while Bopha was still at 3.8°N, 145.2°E, it did grow into a Category 1 typhoon.

Bopha then rapidly gained in intensity. On 1 December, while at 5.8°N, 138.8°E, it had intensified into a C4 Super Typhoon. On 2 December, wind speeds were 259 km/h, those of a C5 Super Typhoon. Notably, this happened while Bopha was at 7.4°N, 128.9°E, closer to the equator than any Category 5 tropical cyclone ever had before. On 3 December, as Bopha interacted with Palau Island, it weakened temporarily into a C3 typhoon before reintensifying back to C5. On 2 December, Bopha entered the Philippine area of responsibility at 8 a.m. local time and was assigned the local name of Pablo.

Bopha crossed the eastern Mindanao coast at about 7.7°N on 4 December at 0445H, the global record proximity to the equator for all C5 tropical cyclones (**Figure 4B**). Average wind speeds and gusts were 185 and 210 km/h, respectively. Many fisher folk at sea were lost, and many coastal dwellers were drowned.

Once onshore, Bopha weakened rapidly as it expended much of its energy in wreaking great havoc. Numerous deaths and severe injuries were attributed to flying trees and debris [29]; however, by far, the greatest cause of death and destruction was the Mayo River debris flow that the typhoon rains generated (**Figure 4C**).

Bopha passed through Mindanao, entered the Sulu Sea, and crossed Palawan Island to enter the West Philippine Sea. There, it reversed course and approached northern Luzon but dissipated before reaching it.

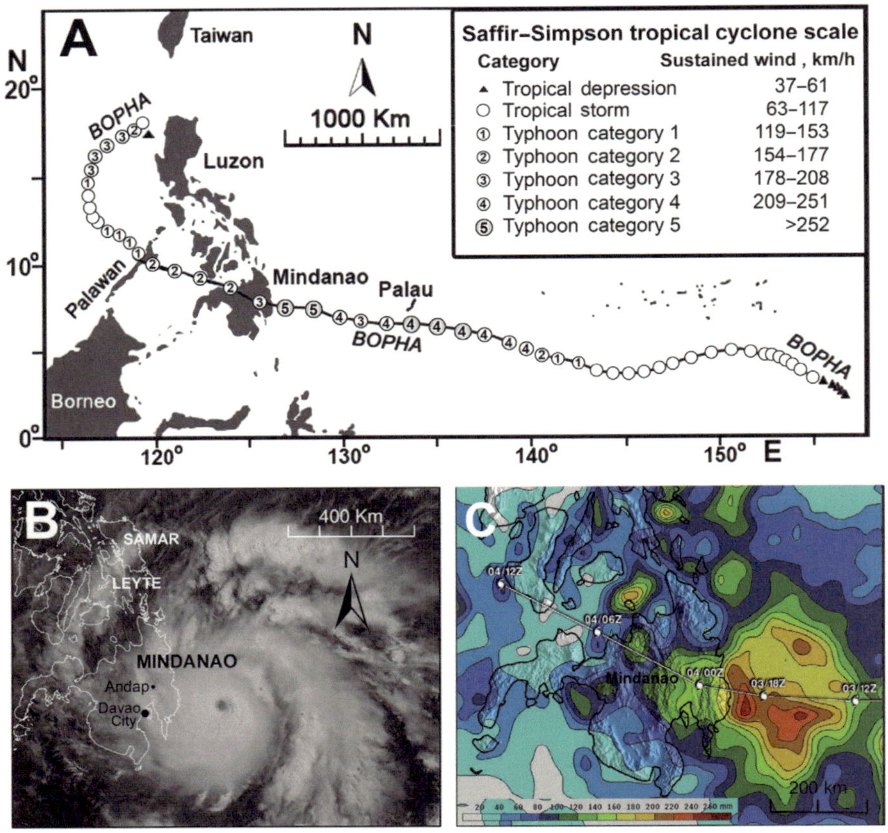

Figure 4. Typhoon Bopha (Pablo). (A) Track and development of the Super Typhoon. New Bataan, Andap, and the Maragusan rain gauge lie beneath the Category 3 icon following the Mindanao landfall. (B) Bopha nearing landfall (modified from [28]). (C) Tropical Rainfall Measurement Mission (TRMM) image from which NASA [28] estimated that Bopha delivered over 240 mm of rainfall near the coast.

The United Nations Office for the Coordination of Humanitarian Affairs (UNOCHA) [30] reported that 1146 Filipinos were killed by Bopha; 834 were still missing, and 925,412 were rendered homeless. It totally or partially destroyed more than 233,000 and caused 1.04 billion U.S. dollars of damage to buildings, crops, and infrastructure. Bopha was the most costly typhoon in Philippine history up to that time—only to be superseded less than a year later in November 2013 when Super Typhoon Haiyan generated the storm-surge that destroyed Tacloban City and devastated widespread areas in the central Philippines.

4. The Mayo River debris flow

The available rain gauge data for Bopha were gathered at Maragusan municipality, 17 km south of Andap (**Figure 5**). Even at that distance, given the Bopha's huge size, these data are good proxies for the rainfall that caused the debris flow. They show that the Mayo River watershed received 120 mm of rain from midnight on 4 December until the flow occurred at 6:30 that morning. It fell as intensely as 43 mm/h, and 4.4 million m^3 were accumulated. These values greatly exceeded the debris-flow initiation thresholds at the Philippine volcanoes Mayon and Pinatubo [31, 32] and Taiwan [33, 34].

After the debris flow began, it was sustained by another 24 mm of torrential rain that fell until 7 a.m., delivering an additional 900,000 m^3 of water. The rainstorm peak in intensity, 52 mm/h, happened at 6:45 a.m. A half-kilometer downstream of the Mayo Bridge (**Figure 1**), the Mamada River discharges into the Kalyawan River; the storm runoff from its 17.7 km^2 watershed, along with the discharge from other Kalyawan tributaries, diluted the debris flow into hyperconcentrated flows that extended 2 km beyond Cabinuangan (**Figure 2B**).

Several geological factors contributed to make the debris flow possible. The Mati Fault in **Figure 1** is a major splay of the Philippine Fault zone, so the rocks of the Mayo watershed have undergone extensive fracturing, making abundant rock debris and facilitating its weathering into soils. Mining and logging has denuded the watershed slopes, facilitating landslides. Bopha's winds uprooted trees on the slopes, exposing soils to storm runoff. Soils are rich in clay, which increases the debris-flow mobility and runout distance [35]. Furthermore, the 2012 debris flow swelled as it easily incorporated the old debris-flow deposits that lay abundantly along its path (**Figure 3B**).

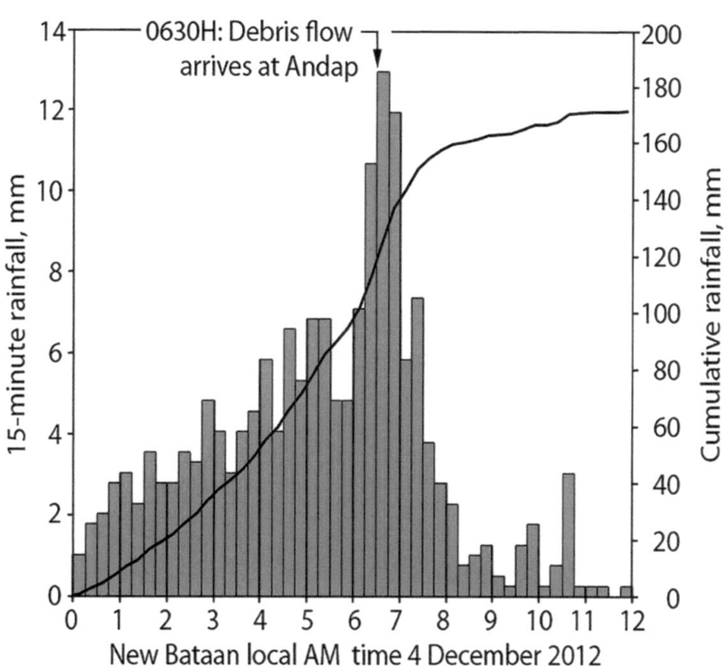

Figure 5. The rainfall that triggered and sustained the Mayo debris flow. Histogram measures the rain that fell during successive 15-minute intervals; 120 mm had accumulated by the time the debris flow hit Andap. Another 24 mm of peak rainfall sustained the flow until 0700H before the storm began to wane.

At about 6:30 a.m., Andap resident Eva Penserga watched in horror as a sturdy concrete bridge 1.5 km upstream of Andap was obliterated by the 16-m high front of a full-fledged debris flow emerging from the Mayo River gorge. A truck on the bridge carrying 30 construction workers was carried away. Several minutes later, surviving Andap residents watched for 5–10 minutes as the debris flow passed through the village.

Tragically, alerts radioed by the government before the catastrophe had urged people to avoid floods at the Andap community center because it stood on high ground. About 200 of them joined the local inhabitants there; 566 people were swept away, 7.5% of the population counted in Andap by the 2010 census.

We calculated a maximum debris-flow velocity of 60 km/h from amateur videos and the length of the debris-flow deposit. Nothing could withstand the main flow, but along its eastern edge, 70 m upstream from the obliterated Andap community center, slower velocities are documented by damage to trees that still survived (**Figure 6**). When a flow of water or debris encounters and rides up an obstruction, the height to which it rises is a measure of its velocity [36]. If all of the kinetic energy of the flow was converted to potential energy as it rose up against the trees, the 1.8 m run-up height h recorded by the highest damage indicates a velocity v of 5.8 m/s, or 21 km/h, from $v = (2gh)^{1/2}$. This is only a minimal value, because the formula takes neither channel roughness nor internal friction into account.

Our satellite imagery, the maps of Bopha after the disaster that we made with LiDAR data, and our field measurements yield a volume of the Andap debris-flow deposit of 25–30 million

Figure 6. Data used to reconstruct the velocity of the debris flow that destroyed Andap. The woman is pointing at the flow level before run-up. Remnants of Andap are in the background; the yellow sign commemorates all the victims by name.

m^3. This ranks it seventh among the largest debris flows of the world (**Table 1**). The width of the deposit is 0.2–1 km wide. The deposits with the greatest thickness of 9 m, in the 500 ha Andap area, includes boulders 16 m in size (**Figure 3C**). Thicknesses decrease downstream to about 0.25 m, and the sediments diminish in size into pebbly, laminated hyperconcentrated-flow sands that cover a 2000 ha area which extends 8 km north of Cabinuangan. In this area, abundant tree trunks and other forest debris (**Figure 7**) that contained many cadavers accumulated in creeks. To impart a sense of the maximal damage, the debris flow wreaked upstream of Cabinuangan, and **Figure 8** presents before and after aerial coverage of the Mayo Bridge and Andap areas.

Figure 7. Hyperconcentrated flows left large tangles of lumber and tree debris with many cadavers at numerous Cabinuangan (central New Bataan) sites.

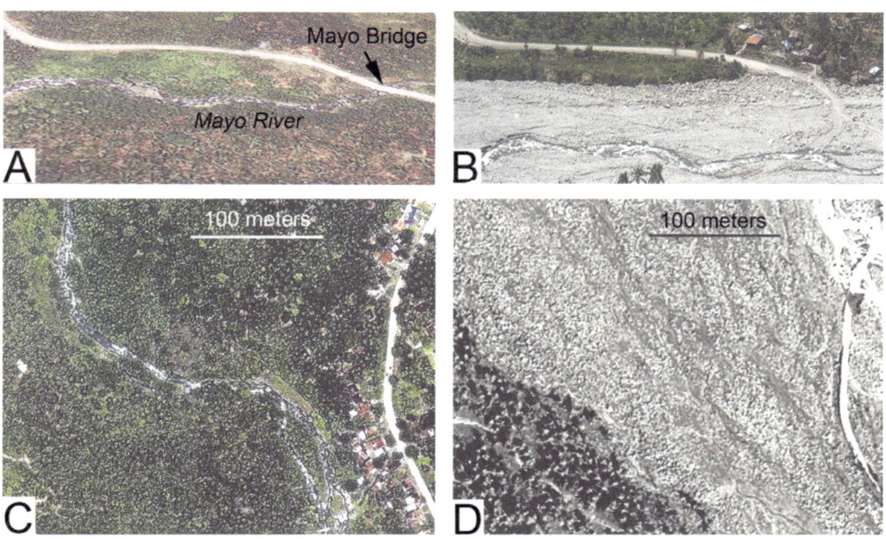

Figure 8. Upstream impacts of the debris flow. (A) Google image of the Mayo River and Mayo Bridge before the debris flow. (B) Post-Bopha air photo of the same area. The large boulder in **Figure 3C** is where the temporary road crosses the stream. (C) Air photo of Andap before Bopha. (D) Same area after the debris flow.

5. The role of Philippine population growth

The global increase in death and damage from natural calamities can be ascribed in part to anthropogenic climate change, but another important reason is the expansion of growing populations into high-risk areas [37]. Landslide disasters are also increasing in developed European and North American nations [35–39], but the trend is especially pronounced in places visited by tropical cyclones [40, 41]. Nowhere is this better exemplified than in Mindanao and by the Andap catastrophe.

The founding of the newer Mindanao settlements including New Bataan was largely driven by rapid population growth [42]. In 1950, the average Filipino farmer cultivated one hectare; this was cut in half by the early 1980s. When New Bataan was settled in 1968, the Philippine population was 36.4 million and was growing 2.98% a year. By July 2018, it had almost tripled, to 106.6 million [43]. A congressional bill filed in 2003 was meant to provide contraception to the poor. The Philippines is predominantly Roman Catholic, and only after strenuous clerical opposition was the bill finally passed in December 2012—coincidentally, the month that Bopha arrived. It must be said that, for all his failings, Rodrigo Duterte is the first Philippine President to take population growth seriously, beginning with his providing contraception to the poor of Davao City when he was its Mayor. Still, the annual growth rate has dropped to 1.72%, but that equates to two million more Filipinos annually. Virtually no areas free of hazards are available to house them.

The Philippine coastal areas, which provide housing and sustenance for two-thirds of the population, are fully developed and increasingly crowded. Metro Manila, the most populated and fastest-growing area, is extracting so much that it is subsiding several centimeters to more than a decimeter annually, losing area to the sea and becoming ever less able to accommodate more people because of worsening floods and tidal incursions [44]. Other rapidly growing Philippine coastal cities including Davao City southwest of Andap (**Figure 4B**) are probably experiencing the same problems.

Real-estate interests are taking advantage of the urgent need for living space by seeking to reclaim 38,272 ha of Philippine coastal areas, including 26,234 ha that comprise virtually the entire near-shore zone of Manila Bay [45]. This, even though rapid subsidence increasingly subjects coastal Metro Manila to storm surges, and the metropolis is overdue for a major earthquake, enhanced ground shaking and liquefaction that would disproportionately damage reclaimed land. Inexorably, other people are seeking living space inland, where natural hazards abound, especially landslides and debris flows. Davao City is a 100-km southwest of Andap (**Figure 4B**). Before leaving to assume the Presidency, Mayor Rodrigo Duterte approved a 200-ha reclamation project for the city [46]. No geological feasibility studies were conducted. The city shares a similar geographic setting with Tacloban City, at the head of a bay. In 2013, a huge storm surge generated by Typhoon Haiyan was funneled up the bay to obliterate much of Tacloban. Davao City is close to segments of the Philippine Fault; offshore earthquakes could similarly funnel tsunamis up to Davao.

6. Will climate change bring more frequent typhoons to Mindanao?

6.1. The historical record of tropical cyclone landfalls in Mindanao

The alluvial fan on which Andap was built, and the ancient debris-flow deposits under New Bataan testify that such flows occurred many times before Super Typhoon Bopha. The youngest of these left the deposits underlying the highway in **Figure 3B**; the sizes of some trees rooted in old deposits (**Figure 3A**) indicate that one event occurred decades or even a century before New Bataan was settled in 1968.

Do these old debris-flow deposits and the Andap disaster merely represent the most recent rare and essentially random Super Typhoons, or will Mindanao and other low-latitude regions suffer from such catastrophes more frequently as the climate changes? Most climatologists [47–52] equate climate change with fewer but more intense tropical cyclones due to rising sea-surface temperatures and atmospheric water vapor. But this says nothing about whether typhoons will hit Mindanao more frequently in the future, even though their history since 1945 suggests as much (**Figure 9A**).

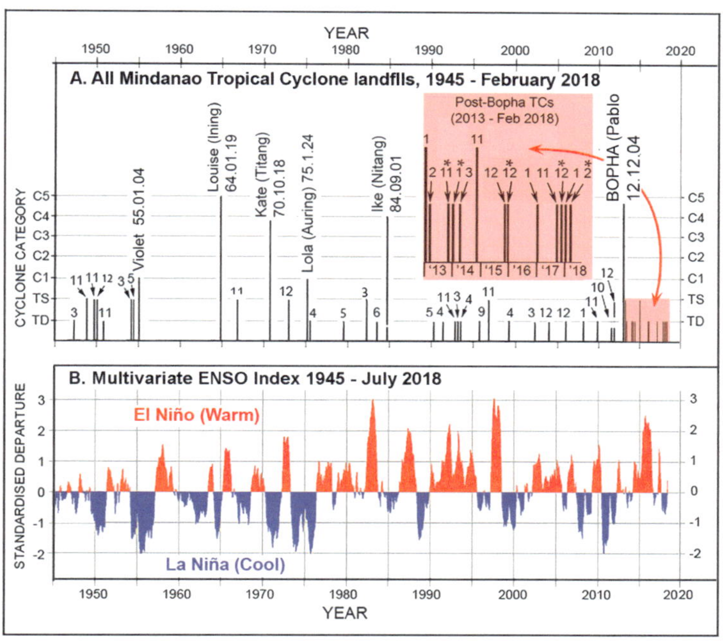

Figure 9. Historical record of Mindanao typhoons. (A) The record from 1945 to February 2018. Cyclone categories: TD = tropical depression; TS = tropical storm; C1–C5 are categories increasing in strength; their respective sustained wind speeds are given in **Figure 4A**. Each typhoon is dated as YY.MM.DD. Number over each TD and TS is the month of occurrence, from 1 January to 12 December. In the insert, asterisked TDs caused fatalities. (B) The record of all El Niño and La Niña events since 1945. From [58, 59].

The literature regarding the frequency of West Pacific typhoons yields little insight pertinent to Mindanao. Two reports [53, 54] exemplify the problem; the significant multi-year fluctuations they reported are not manifested in Mindanao. Neither study included tropical depressions among their data because these are harder to define and are more ephemeral, and thus they ignored most of the Mindanao occurrences. Landfalls are easy to locate; however, tropical depressions can be lethal on Mindanao. For example, 5 of the 13 that came after Bopha (**Figure 9A** insert) caused fatalities. In addition to ignoring tropical depressions, the data set utilized by reference [54], which is entitled "Inactive Period of Western North Pacific Tropical Cyclone Activity in 1998 –2011," was limited to tropical cyclones of the June -October main typhoon season. Of the 45 that made landfall on Mindanao since 1945, only 5 came in that season: Typhoons Kate in October 1970 and Ike in September 1984, and 3 of the 11 tropical depressions. For Mindanao, unlike for the northwest Pacific as a whole, 1998–2011 was not at all a slack cyclone period.

Until 1990, tropical cyclones rarely and sporadically made landfall on Mindanao because it was in the ephemeral southern fringe of the northwest Pacific typhoon track. Since the U.S. Navy Joint Typhoon Warning System began to archive northwest Pacific tropical cyclones in 1945, only 34 visited Mindanao by February 2012 [55]. After Bopha, another 13 had arrived by February 2018 (**Figure 9A**). These 47 landfalls are incontrovertible, and our search for what the future might hold begins with them.

During 30 of the 45 years from 1945 to 1990, including the 8 years from 1956 to 1963, not one tropical depression visited Mindanao. Most of the rare Mindanao tropical cyclones were weak: six tropical depressions and eight tropical storms. Since 1955, five typhoons did make landfall on Mindanao before Bopha, although Louise and Ike barely crossed northernmost Mindanao. In 1970, C4 Kate passed 45 km south of New Bataan, which experienced heavy rain and floods but not much wind. Only five tropical cyclones of all categories arrived during the northwest Pacific peak typhoon season of June through October, although these included Kate in October 1970 and Ike in September 1984. A total of 36 came during the off season: 18 in March through June and 18 during November through January.

In the period from 1945 to 1989, Mindanao tropical cyclones occurred only every 2.5 years on average. Then, from 1990 to 2017, they began arriving roughly once a year. It is also troublesome that Mindanao has recently begun to suffer lethal cyclones in consecutive years. In December 2011, a year before Bopha, Tropical Storm Washi killed 1268 people in Cagayan de Oro City, on the northern Mindanao coast 180 kilometers from New Bataan [56, 57]. Two months earlier, a tropical depression arrived in Mindanao, making 2011 only the fifth year since 1945 for the island to receive two tropical cyclones. In 2014, Mindanao experienced Tropical Depression Lingling and Tropical Storm Jangmi. Last, in the 5 years since Bopha, 2 tropical storms and 11 tropical depressions have visited Mindanao—so frequently that they had to be plotted as an insert in **Figure 9A**.

The increasing frequency of Mindanao storms since 1990, although alarming, cannot be ascribed simply to the climate change from global warming. We must consider whether these changes are tied to multi-annual and multi-decadal fluctuations in western North Pacific sea-surface temperatures.

6.2. El Niño-Southern Oscillation (ENSO) and typhoon frequency

ENSO is the complex result of ocean-atmosphere interactions that are best expressed by fluctuating sea-surface temperatures in the central and eastern equatorial Pacific (**Figure 9B**), from warmer during El Niño to cooler during La Niña periods [58, 59]. During an El Niño, atmospheric pressures are high over the western Pacific and low in the eastern Pacific, and the situation reverses during a La Niña. Both phases occur every 3–5 years; typically, El Niño episodes last 9 months to a year and La Niñas as long as 3 years [60].

Tropical cyclones tend to form farther east, are more widely dispersed, and curve northward during El Niño episodes, thus are rarer in the Philippines. During La Niñas, their tendency is to start farther west, remain below 23°N, and take more westward courses, and so they make more frequent Philippine landfalls, most during a main September–November season [48, 51, 61–63]. Except for their tendency to arrive later than November, all the typhoons before Bopha that made Mindanao landfalls since 1945 fit that pattern by occurring during La Niñas (**Figure 9**). Bopha came either during a weak La Niña [64] or a weak El Niño [65]. From 1945 to 2018, the weaker storms and depressions that visited Mindanao showed no marked preference between El Niño and La Niña episodes, although they tended to occur during La Niñas from 1945 to 1975 and 1996 to 2012 and during El Niños from 1979 to 1995 and 2013 to 2018.

A major analysis [66] has predicted that global warming will increase the historical average frequency of extreme La Niñas from one every 23 years to one every 13 years. Three effects of global warming are blamed: first, the western North Pacific region of insular seas and islands that includes the Philippines is expected to warm more quickly than the central Pacific; second, the temperature gradients in the surface waters of the tropics will increase; and last, extreme El Niños, which will occur more frequently, are usually followed by extreme Niñas [67]. Given the tendency of typhoons to make landfalls on the Philippines more frequently during La Niñas, the country including Mindanao should expect greater future storminess.

Another cycle of sea-surface temperature, the Pacific Decadal Oscillation (PDO), is so-called because its periods last for two or three decades [68, 69]. This cyclicity is most distinctly expressed in the northern Pacific but is not clearly manifested in the tropics, although one study [70] attributes an interdecadal variation in May-June rainfall over southern China to a complex interplay between the PDO and the ENSO. The PDO does not correlate with Mindanao landfalls, or even with the frequency of all northwest Pacific typhoons.

6.3. Other expected behaviors of future typhoons

Seven prominent climate scientists who reviewed [52] the great body of research on how climate change might be affecting tropical-cyclone activity explain that its many conflicting results arise from great variations in cyclone frequencies and intensities, as well as serious lacks in the quantity and quality of the records. The authors are not sure that the observed changes exceed the variability due to natural causes, but predict that by 2100 the averaged *frequencies* of all tropical cyclones will decrease 6–34%, They also believe, however, that *intensities* will increase 2–11% by century's end because, although the frequency of all tropical cyclones is expected to decrease, the most intense ones will become more frequent. Importantly, the review predicts a 20% increase of rainfall within 100 kilometers of storm centers, which would generate larger debris flows. This increase is ascribed [71] to anthropogenic warming, which weakens the summertime winds that carry the tropical cyclones along. Already, their translation speeds decreased globally by 10% from 1949 to 2016. This slowing enhances the amount of time they have to take up water vapor from the ocean and deliver rain when their centers reach land.

In short, the record of increasingly frequent landfalls on Mindanao may or may not indicate that more frequent typhoon disasters will happen there in the future, although recent reports [66, 67] strongly imply as much. Low-latitude areas, however, are given short shrift by most meteorological and climatologic analyses. We urgently need to understand how anthropogenic global warming is changing tropical-cyclone behavior in subequatorial regions because so many people live in them.

7. A new Philippine catalog of alluvial fans and their associated debris-flow hazards

Following our study of the Andap disaster, Project NOAH used high-resolution digital terrain models to identify and catalog all Philippine alluvial fans, by analyzing geomorphic features, slopes, gradients, and stream networks nationwide. The catalog is accessible online for free in the NOAH portal [72].

More than 1200 alluvial fans were identified, and communities that might be affected by their debris flows are being educated about the hazard. In October 2015, Typhoon Koppu (Lando) generated devastating debris flows on alluvial fans in Nueva Ecija province, but the vulnerable communities were warned and evacuated, and so no one was killed [73]. Later that year, Typhoon Melor (Nona) also triggering massive debris flows in Mindoro Island, burying or sweeping away houses and infrastructure in several communities situated on alluvial fans. Again, timely warnings and evacuations prevented the loss of life [74].

8. Other climate-related hazards in the Philippines and Mindanao

Future fluctuations between extreme El Niños and La Niñas pose other threats. Philippine rainfall is modulated by ENSO; El Niños bring droughts, and La Niñas cause excessive rainfall [75, 76]. Rock debris accumulates on slopes during a protracted El Niño drought; the succeeding La Niña episode brings heavy downpours that mobilize the accumulated material into landslides and debris flows. Another serious hazard associates with ENSO is forest fires: rainy La Niña episodes promote strong vegetation growth that a succeeding El Niño drought dries out and renders inflammable.

Mindanao has 21 active and potentially active volcanoes [77]. They are tourist attractions, producing geothermal energy, some are actively mined, and many support large agricultural populations. These volcanoes still lack thorough study and monitoring instrumentation, and similar to the situation at Pinatubo Volcano on Luzon Island before its catastrophic 1991 eruption, their populations are unfamiliar with eruptions and lahars. Any major eruption will eventually be followed by a large typhoon and lahars. The larger Mindanao volcanoes, being structurally and mechanically weak [78], do not need to erupt to undergo debris flows. All that is needed to trigger them would be exceptionally strong rainstorms in their vicinities.

9. Conclusions

Bopha formed, became Category 5 Super Typhoon, and made landfall closer to the equator than any C5 tropical cyclone ever had before. More than 120 mm of rain fell on the Mayo River watershed in only 7 hours. A catastrophic debris flow it generated devastated *Barangy* Andap and killed 566 of its inhabitants. We measured its deposit as a dry volume of 30 million m^3, making it the seventh largest globally.

Debris flows are remarkably poorly understood in the Philippines. This is especially true in Mindanao because it is located in the southern fringe of the typhoon track of the northwest Pacific and has rarely experienced typhoons and the debris flows they generate. This lack of experience is a main cause of the loss of life in Andap.

New Bataan and Andap were established in 1968 by people who did not understand the nature of the ancient debris-flow deposits on which they were building and the hazard that produced them. This was still the case when Bopha approached: government authorities broadcast the fatal advice for people to avoid flooding on the high ground at Andap, which was sitting on the Mayo River alluvial fan. The lack of understanding about debris flows persisted after the disaster; government scientists assigned to explain the tragedy and select relocation sites for the displaced people called it a "flash flood" [78].

New Bataan and Andap were settled in the late 1960s because of rapid population growth. The population continues to explode and has to occupy areas vulnerable to natural hazards. The lesson of Andap and numerous other recent disasters is that new settlements must not be established before the hazards that threaten them have been properly evaluated. But this is a daunting requirement, because few, if any, safe sites remain unoccupied.

Whether or not Mindanao will experience more frequent typhoons and debris flows is an urgent question that is very difficult to answer. In 1945, Western North Pacific tropical cyclones began to be archived accurately; by 1990, the frequency of Mindanao landfalls had doubled. Learning whether this is caused by anthropogenic global warming is complicated by deficiencies in the quantity and quality of the archived data and by the irregularities in the ENSO climatic rhythms. For Mindanao, the problem is especially difficult because most of its tropical cyclones do not arrive in the main typhoon season of July through October, and most are only tropical depressions, which most climatologists and meteorologists do not include as data for their models.

Philippine typhoons occur most frequently during La Niña episodes, and from July to October, in Mindanao, however, they arrive during the off season from November to June. Extreme El Niños and La Niñas are expected to succeed each other more frequently. This is an excellent example of how Earth systems, which are kept in balance by numerous interacting phenomena, oscillate vigorously when they are disturbed. Global warming is a continuing and accelerating disturbance that prevents returns to equilibria. Mindanao and the entire Philippine nation urgently need to prepare their populations for more frequent hazards, including floods, storm surges, landslides, debris flows, and forest fires.

A developing country like the Philippines has limited resources for hazard-mitigation measures. Philippine society is intensely focused on the family, and so the best and least expensive governmental approach is to provide every family with good, easily-accessible information, so it can develop its own emergency plans.

Project NOAH's mandate tasks are to evaluate the nation's numerous natural hazards, to educate each community about the hazards that threaten it, and to advise them how to respond when a threat materializes. Our study of the Mayo debris flow motivated us to identify more than 1200 Philippine alluvial fans and to prepare the communities that its debris flows may affect. This work has already helped to save lives from major debris flows in 2015.

Acknowledgements

This work was funded by the Philippine Department of Science and Technology (DOST) and the Volcano Tectonics laboratory of the National Institute of Geological Sciences at the University of the Philippines (U.P.). LiDAR data covering the New Bataan area were provided by the U.P. Training Center for Applied Geodesy and Photogrammetry. DOST's Advanced Science and Technology Institute and the Philippine Atmospheric, Geophysical and Astronomical Services Administration provided rainfall data. We thank Congresswoman M. C. Zamora for logistical support and Thomas Pierson for the information about debris-flow mechanics.

Conflict of interest

We have no conflict of interest to declare.

Author details

Kelvin S. Rodolfo[1]*, A. Mahar F. Lagmay[2], Rodrigo C. Eco[3], Tatum Miko L. Herrero[4], Jerico E. Mendoza[5], Likha G. Minimo[6], Joy T. Santiago[5], Jenalyn Alconis-Ayco[7], Eric C. Colmenares[8], Jasmine J. Sabado[9] and Ryanne Wayne Serrado[10]

*Address all correspondence to: krodolfo@uic.edu

1 University of Illinois at Chicago, Chicago, Illinois, USA

2 UP Resilience Institute, University of the Philippines, Diliman, Quezon City, Philippines

3 National Institute of Geological Sciences, University of the Philippines, Diliman, Quezon City, Philippines

4 GEOMAR—Helmholtz Centre for Ocean Research Kiel, Germany

5 Nationwide Operational Assessment of Hazards Center, University of the Philippines, Quezon City, Philippines

6 Department of Geological Sciences, University of Canterbury, Christchurch, New Zealand

7 Development Network Consulting Services, University of the Philippines, Quezon City, Philippines

8 KNPN Technologies, Davao City, Philippines

9 Development Academy of the Philippines, Ortigas Center, Pasig, Philippines

10 Clariden Holdings, Inc., Mandaluyong City, Philippines

References

[1] Rodolfo KS, Lagmay AMF, Eco RC, Herrero TM, Mendoza JE, Minimo LG, et al. The December 2012 Mayo River debris flow triggered by Super Typhoon Bopha in Mindanao, Philippines: Lessons learned and questions raised. Natural Hazards and Earth System Sciences. 2016;**16**:2683-2695. DOI: 10.5194/nhess-16-2683-2695

[2] Ea C, Lebumfacil MA, Silveron JM, Juson B. Proposal to Declare Certain Areas in New Bataan Identified as Highly Susceptible to Landslide and Flooding Geohazard [Internet]. 2013. Available from: https://www.academia.edu/5054503/POLICY_PROPOSAL_TO_DECLARE_CERTAIN_AREAS_IN_NEW_BATAAN_IDENTIFIED_AS_HIGH_TO_LANDSLIDE_AND_FLOODING_AS_NO_BUILD_ZONE [Accessed: November 24, 2015]

[3] 1970 Census of Population and Housing Final Report for Davao del Norte. National Census and Statistics Office. Manila; National Economic and Development Authority; 1974. p.1

[4] National Statistics Office. Total Population by Province, City, Municipality and Barangay as of May 1, 2010. Compostela Valley, Census of Population and Housing [Internet].

2010. Available from: http://web0.psa.gov.ph/sites/default/files/attachments/hsd/press-release/Davao.pdf [Accessed: November 24, 2015]

[5] MGB (Mines and Geosciences Bureau. List of Susceptible Areas to Mass Movements in Region XI as Covered Under the National Geohazard Mapping Program [Internet]. 2009. Available from: http://mgbxi.org/list-of-susceptible-areas-to-mass-movements-in-region-xi/ [Accessed: November 24, 2015]

[6] Takahashi T. Debris flows. Annual Review of Fluid Mechanics. 1981;**13**:57-77

[7] Hutter K, Svendsen B, Rickenmann D. Debris flow modeling: A review. Continuum Mechanics and Thermodynamics. 1994;**8**:1-35

[8] Iverson RM. The physics of debris flows. Reviews of Geophysics. 1997;**35**(3):245-296

[9] Iverson RM, Reid ME, LaHusen RG. Debris-flow mobilization from landslides. Annual Review of Earth and Planetary Sciences. 1997;**25**:85-138

[10] Jakob MA. Size classification for debris flows. Engineering Geology. 2005;**79**:151-161

[11] Vallance JW. Lahars. In: Sigurdsson H et al., editors. Encyclopedia of Volcanoes. San Diego: Academic Press; 2000. pp. 973-995

[12] Rodolfo KS. The hazard from lahars and jokulhlaups. In: Sigurdsson H et al., editors. Encyclopedia of Volcanoes. San Diego: Academic Press; 2000. pp. 973-995

[13] Lagmay AMF, Rodolfo KS, Siringan FP, Uy H, Remotigue C, Zamora P, et al. Geology and hazard implications of the Maraunot notch in the Pinatubo Caldera, Philippines. Bulletin of Volcanology. 2007;**69**:797-809

[14] Schuster R, Salcedo DA, Valenzuela L. Overview of catastrophic landslides on South America in the twentieth century. Reviews in Engineering Geology. 2002;**15**:1-34

[15] Remotigue C. The 5-6 October 1993 lahar event on the Bucao River, Zambales: Observed dynamics, effects and telemetered data [M.S. thesis]. Diliman, Quezon City, Philippines: National Institute of Geological Sciences, University of the Philippines; 1995

[16] Haeberli W, Huggel C, Kääb A, Zgraggen-Oswald S, Polkvoj A, Galushkin I, et al. The Kolka-Karmadon rock/ice slide of 20 September 2002: An extraordinary event of historical dimensions in North Ossetia, Russian Caucasus. Journal of Glaciology. 2004;**50**(171):33-546

[17] Plafker G, Ericksen GE. Nevados Huascaran avalanches, Peru. Rockslides and Avalanches. 1978;**1**:277-314

[18] Wieczorek GF, Larsen MC, Eaton LS, Morgan BA, Blair J. Debris-flow and flooding hazards associated with the December 1999 storm in coastal Venezuela and strategies for mitigation. U.S. Geological Survey Open File Report; 2001. No. 01-0144

[19] Paguican EMR, Lagmay AMF, Rodolfo KS, Rodolfo RS, Tengonciang AMP, Lapus MR, et al. Extreme rainfall-induced lahars and dike breaching, 30 November 2006, Mayon Volcano, Philippines. Bulletin of Volcanology. 2009;**71**:845-857

[20] Pierson TC. Initiation and flow behavior of the 1980 Pine Creek and Muddy river lahars, Mount St. Helens, Washington. Geological Society of America Bulletin. 1985;**96**(8):1056-1069

[21] Pierson TC, Scott KM. Downstream dilution of a lahar: Transition from debris flow to hyperconcentrated stream flow. Water Resources Research. 1985;**21**:1511-1524

[22] Smith GA, Lowe DR. Lahars: Volcano hydrologic events and deposition in the debris flow—Hyperconcentrated flow continuum. In: Fisher RV, Smith GA, editors. Sedimentation in Volcanic Settings. Tulsa; SEPM Special Publication. 1991;**45**:59-70

[23] Pierson TC. An empirical method for estimating travel times for wet volcanic mass flows. Bulletin of Volcanology. 1998;**60**:98-109

[24] Rickenmann D. Empirical relationships for debris flows. Natural Hazards. 1999;**19**:47-77

[25] Gallino GL, Pierson TC. Polallie Creek debris flow and subsequent dam-break flood of 1980, East Fork Hood River basin, Oregon. U.S. Geological Survey Water-Supply Paper. 1985:84-578

[26] Savage SB, Lun CKK. Particle size segregation in inclined chute flow of dry cohesionless granular solids. Journal of Fluid Mechanics. 1988;**189**:311-335

[27] Pierson TC. Distinguishing between debris flows and floods from field evidence in small watersheds. U.S. Geological Survey Fact Sheet. 2005:2004-3142

[28] NASA Earth Observatory. Bopha Makes Landfall [Internet]. 4 December 2012. Available from: http://earthobservatory.nasa.gov/NaturalHazards/ view.php?id=79892 [Accessed: June 21, 2018]

[29] National Disaster Risk Reduction and Management Council (NDRRMC). Situation Report No. 38 re effects of "Typhoon PABLO" (Bopha) [Internet]. December 2012. Available from: http://reliefweb.int/report/philippines/ndrrmc-update-sitrep-no-38-re-effects-typhoon-pablo-bopha [Accessed: June 21, 2018]

[30] United Nations Office for the Coordination of Humanitarian Affairs (UNOCHA) Philippines. Typhoon Bopha Situation Report No. 19 (final) [Internet]. 12 February 2013. Available from: https://reliefweb.int/report/philippines/typhoon-bopha-situation-report-no-19-12-february-2013 [Accessed: June 21, 2018]

[31] Rodolfo KS, Arguden AT. Rain-lahar generation and sediment delivery systems at Mayon Volcano, Philippines. In: Fisher RV, Smith GA, editors. Sedimentation in Volcanic Settings. Vol. 45. Society of Economic Mineralogy and Paleontology Special Publications. 1991. pp. 71-87

[32] Westen CJV, Daag AS. Analysing the relation between rainfall characteristics and lahar activity at Mount Pinatubo. In: Earth Surface Process and Landforms. Vol. 30. 2005. pp. 1663-1674

[33] Guzzetti F, Peruccacci S, Rossi M, Stark CP. The rainfall intensity-duration control of shallow landslides and debris flows. Landslides. 2008;**5**:3-17

[34] Huang C-C. Critical rainfall for typhoon-induced debris flows in the Western Foothills, Taiwan. Geomorphology. 2013;**185**:87-95

[35] Costa JE, editor. Developments and Applications of Geomorphology. Berlin: Springer-Verlag; 1984. pp. 268-317

[36] Arguden AT, Rodolfo KS. Sedimentologic and dynamic differences between hot and cold laharic debris flows of Mayon Volcano, Philippines. Geological Society of America Bulletin. 1990;**102**:865-876

[37] Huppert HE, Sparks S. Extreme natural hazards: Population growth, globalization and environmental change. Philosophical Transactions of the Royal Society A. 2006;**364**:1875-1888

[38] Cascini L, Bonnard C, Corominas J, Jibson R, Montero-Olarte J. Landslide hazard and risk zoning for urban planning and development. In: Hungr O et al., editors. Landslide Risk Management. Boca Raton. CRC Press; 2005. pp. 199-235

[39] Di Martire D, De Rosa M, Pesce V, Santangelo MA, Calcaterra D. Landslide hazard and land management in high-density urban areas of Campania region, Italy. Natural Hazards and Earth System Sciences. 2012;**12**:905-926

[40] Lari S, Frattini P, Crosta GB. Local scale multiple risk assessment and uncertainty evaluation in a densely urbanised area (Brescia, Italy). Natural Hazards and Earth System Sciences. 2012;**12**:3387-3406

[41] Weinkle J, Maue R, Pielke R Jr. Historical global tropical cyclone landfalls. Journal of Climate. 2012;**13**:4729-4735

[42] Dolan RE, editor. Philippines: A Country Study. Area Handbook Series 1993; 550.72) [Internet]. Federal Research Division, Library of Congress, US Government Printing Office, 384. Available at: http://countrystudies.us/philippines/ [Accessed: June 23, 2018]

[43] World Population Review. Philippine Population. [Internet]. 2018-06-14. Available from: http://worldpopulationreview.com/countries/philippines/ [Accessed: June 15, 2018]

[44] Rodolfo KS, Siringan FB. Global sea-level rise is recognised, but flooding from anthropogenic land subsidence is ignored around northern Manila Bay, Philippines. Disasters. 2006;**30**(1):118-139

[45] Rodolfo KS. On the geological hazards that threaten existing and proposed reclamations of Manila Bay. Philippine Science Letters. 2014;**7**(1):228-240

[46] Padillo MM. Coastline-port development project endorsed to NEDA [Internet]. Business World. 10 April 2017. Available from: http://www.bworldonline.com/content.php?section=Nation&title=coastline-port-development-project-endorsed-to-neda&id=143535 [Accessed: July 14, 2018]

[47] Webster PJ, Holland GJ, Curry JA, Chang H-R. Changes in tropical cyclone number, duration, and intensity in a warming environment. Science. 2005;**309**:1844-1845

[48] Emanuel K. Increasing destructiveness of tropical cyclones over the past 30 years. Nature. 2005;**436**(4):686-688

[49] Bengtsson L, Hodges KI, Esch M, Keenlyside N, Kornblueth L, Luo JJ, et al. How may tropical cyclones change in a warmer climate? Tellus A. 2007;**59**:539-561

[50] Elsner JB, Kossin JP, Jagger TH. The increasing intensity of the strongest tropical cyclones. Nature. 2008;**4554**:92-95

[51] Emanuel K, Sundararajan R, Williams J. Hurricanes and global warming: Results from downscaling IPCC AR4 simulations. Bulletin of the American Meteorological Society. 2008;**89**:347-367

[52] Knutson TR, McBride JL, Chan J, Emanue K, Holland G, Landsea C, et al. Tropical cyclones and climate change. Nature Geoscience. 2010;**3**(3):157-163

[53] Matsuura T, Yumoto M, Iizuka S. A mechanism of interdecadal variability of tropical cyclone activity over the northwest Pacific. Climate Dynamics. 2003;**2**:105-117

[54] Liu KSJ, Chan CL. Inactive period of western north pacific tropical cyclone activity in 1998-2011. Journal of Climate. 2013;**26**:2614-2630

[55] Unisys Weather. Hurricane/Tropical Data for Western Pacific [Internet]. 2012. Available from: weather.unisys.com/hurricane/w_pacific [Accessed: June 02, 2012]

[56] Ramos BT. Final Report on the Effects and Emergency Management re Tropical Storm "Sendong" (Washi), National Disaster Risk Reduction and Management Center [Internet]. 2013. Available at: http://www.ndrrmc.gov.ph/attachments/article/358/Final%20Report%20re%20TS% [Accessed: November 24, 2016]

[57] Manila Observatory. Tropical Storm Sendong [Internet]. 2011. Available from: www.observatory.ph/cgi-bin/ publications/pubdetails.py?p=137 [Accessed: June 04, 2015]

[58] Trenberth KE. The definition of El Niño. Bulletin of the American Meteorological Society. 1997;**78**(12):2771-2777

[59] Wolter K, Timlin MS. El Niño/Southern Oscillation behaviour since 1871 as diagnosed in an extended multivariate ENSO index (MEI.ext). International Journal of Climatology. 2011;**31**:1074-1087

[60] NOAA Climate Prediction Center. ENSO FAQ [Internet]. How Often do El Niño and La Niña Typically Occur? 2014. Available from: http://www.cpc.noaa.gov/products/analysis_monitoring/ensostuff/ensofaq.shtml #HOWOFTEN [Accessed: November 24, 2016]

[61] Wang B, Chan J-CL. How strong ENSO events affect tropical storm activity over the northwest Pacific. Journal of Climate. 2002;**15**:1643-1658

[62] Wu MC, Chang WL, Leung WM. Impacts of El Niño-Southern Oscillation events on tropical cyclone landfalling activity in the northwest Pacific. Journal of Climate. 2004;**17**(6):1419-1428

[63] Zhang W, Graf HF, Leung Y, Herzo M. Different El Niño types and tropical cyclone landfall in East Asia. Journal of Climate. 2012;**25**(19):6510-6523

[64] NOAA Climate Prediction Center. Cold and Warm Episodes by Season [Internet]. Available from: www.cpc.ncep.noaa.gov/products/analysis_monitoring/ensostuff/ensoyears.shtml [Accessed: November 24, 2016]

[65] National Oceanic and Atmospheric Administration, Earth System Research Laboratory. Multivariate ENSO Index [Internet]. 2018. Available at: https://www.esrl.noaa.gov/psd/enso/mei/ [Accessed: July 12, 2018]

[66] Cai W, Wang G, Santoso A, McPhaden MJ, Wu L, Jin F-F, et al. Increased frequency of extreme La Niña events under greenhouse warming. Nature Climate Change. 2015;**5**:132-137

[67] Cai W, Borlace S, Lengaigne M, van Rensch P, Collins M, Vecchi G, et al. Increasing frequency of extreme El Niño events due to greenhouse warming. Nature Climate Change. 2014;**4**:111-116

[68] Mantua NJ, Hare SR. The Pacific Decadal Oscillation. Journal of Oceanography. 1997; **58**(1):35-44

[69] Joint Institute for the Study of the Atmosphere and Ocean (JISAO), University of Washington. The Pacific Decadal Oscillation, (PDO) [Internet]. 2017. Available from: http://research.jisao.washington.edu/pdo/ [Accessed: July 16, 2018]

[70] Chan JCL, Zhou W. PDO, ENSO and the early summer monsoon rainfall over south China. Geophysical Research Letters. 2005;**32**. ID L08810. 5 pp

[71] Kossin JP. A global slowdown of tropical-cyclone translation speed. Nature. 2018;**558**:104-107

[72] Aquino D, Ortiz IJ, Salvosa S, Timbas NL, Llanes F, Ferrer PK, Atlas of Alluvial Fans in the Philippines. National Institute of Geological Sciences, University of the Philippines [Internet]. 2014. Available from: https://center.noah.up.edu.ph/ [Accessed: July 04, 2018]

[73] Eco R, Herrero TM, Llanes F, Briones , Escap C, Sulapas JJ, Bongabon, Gabaldon and Laur, Nueva Ecija debris flows triggered by Typhoon Koppu [Internet]. NOAH Open File Reports. 2015;**4**:36-43. Available from: http://blog.noah.dost.gov.ph/2015/12/22/bongabon-gabaldon-and-laur-nueva-ecija-debris-flows-triggered-by-typhoon-koppu [Accessed: July 16, 2018]

[74] Llanes FV, Rodolfo KS, Escape CM, Felix R, Ortiz IJ. The December 2015 debris flows triggered by Typhoon Nona in Baco, Oriental Mindoro, Philippines. Geophysical Research Abstracts. 2017;**19**:EGU2017-1483

[75] Lyon B, Cristi H, Verceles ER, Hilario FD, Abastillas R. Seasonal reversal of the ENSO rainfall signal in the Philippines. Geophysical Research Abstracts. 2006;**33**:L24710. DOI: 10.1029/2006GL028182

[76] Philippine Institute of Volcanology and Seismology (PHIVOLCS). Potentially Active Volcanoes [Internet]. 2008. Available from: http://www.phivolcs.dost.gov.ph/index.php?option=com_content& view=article&id=60:potentially-active&catid=55&Itemid=115

[77] Herrero TM. The implications of the structure and morphology of volcanic edifices in the development of mass movements in the Philippine setting [thesis]. Clermont-Ferrand: Université Blaise Pascal, Laboratoire Magmas et Volcans; 2014

[78] Mines and Geosciences Bureau (MGB). Progress report on geohazard mapping of Brgy. Andap and Cabinuangan, New Bataan, Province of Compostela Valley [Internet]. December 2012. Available from: http://mgbxi.org/wp-content/uploads/2013/05/Bgry.-Andap-and-Cabinuangan-Geohazard-FINAL-Report-2.pdf [Accessed: January 06, 2014]

Chapter 7

Observed and Projected Reciprocate Effects of Agriculture and Climate Change: Implications on Ecosystems and Human Livelihoods

Zenebe Mekonnen

Additional information is available at the end of the chapter

http://dx.doi.org/10.5772/intechopen.79118

Abstract

The objective of this chapter is to review, from several literatures, the contribution of agriculture to climate change and the reciprocal effects of climate change on agriculture and the general consequent implications on human livelihoods and ecosystems. Human activities have already had a discernible impact on the earth's climate leading to growing evidence of observable impacts of climate change on physical and biological systems. In no doubt, agriculture provides the world population of 7 billion with the food that we all eat every day. In addition, 1.4 billion people work in agriculture and more than 2.5 billion people sustain their livelihood on agriculture. But agriculture is one of the contributors of greenhouse gases to climate change and climate change affects agriculture in return. When the global mean temperature change increases beyond 3.5°C, most of the species will have very few suitable areas for their survival and will become extinct. Several hundred million people are seriously affected by climate change today, with several hundred thousand annual deaths. Human impacts of climate change include scarcity of freshwater resources, weather-related disasters, food insecurity due to agricultural loss, migration, and displacement due to loss of settlements. These recalled nations to limit their GHG emission, ensure sustainable ecosystem, food production, and economic development so as to calm down the impacts of climate change.

Keywords: adaptation, ecosystems, greenhouse gases, livelihoods, mitigation

1. Introduction

Human activities have already had a discernible impact on the Earth's climate leading to growing evidence of observable impacts of climate change on physical and biological systems [1, 2]. Due to their limited adaptive capacities in technology and affluence as well as high natural resource-dependent livelihoods, it is the least developing countries that are particularly vulnerable to climate change impacts [3, 4]. At recent times, however, other countries in the mid- to high-latitudes have also experienced significantly higher rates of recent warming, and, in the northern hemisphere, such regions have also experienced an increase in heavy precipitation events [2, 5].

In no doubt, agriculture provides the world population of 7 billion with the food that we all eat every day. In addition, 1.4 billion people work in agriculture and more than 2.5 billion people sustain their livelihoods on agriculture [6–8]. Irrespective of all these, intensive agricultural practices have impacted global climate change. It is not only just the actual farming that made intensive agriculture so detrimental, but also land-use changes for its investment—say, deforestation which releases CO_2 as well as increases the surface albedo thereby enhance atmospheric warming [8]. For instance, continuing deforestation, mainly in tropical regions, is currently thought to be responsible for annual emissions of 1.1–1.7 billion tonnes of carbon per year [8].

Agricultural productions need to be increased to accommodate a growing population with reduced emissions of the greenhouse gases (GHGs): carbon dioxide, methane and nitrous oxide [9]. On the other hand, it is becoming apparent that climate change was adversely affected and will continue to affect socio-economic sectors including water resources, agriculture, forestry, fisheries, human settlements, ecological systems and human health in many parts of the world. Developing countries are taking the lion's share of these adverse impacts of climate change and are the most vulnerable [5, 10] due to their low affluence and adaptive capacity to rebuild from climatic shocks. As described by Tol [11], one cannot have cheap energy, beef, mutton, dairy or rice without carbon dioxide emissions. However, employing sustainable practices of agriculture, like organic agriculture, have huge potential to help in the fight against climate change as they can sequester as much as 7000 pounds of carbon dioxide per acre per year [12, 13]. Rising temperatures and changing rainfall patterns had affected the kinds of crops that could have grown in a particular place, with effects unevenly distributed across the world [14]. Climate change was not only affected agriculture but also affected many aspects of human society and the natural world [2]. For instance, climate change had already transformed and will continue to transform ecosystems on an unexpected scale [2, 13–16].

"The linkage from causes to human impact of climate change [17] would surpass through four processes. First, increased emission of GHG has caused climate change; second, climate change has brought significant effects on rising sea surface temperature, sea level, ocean acidification, change in local rainfall and river run off patterns, high species extinction rate, loss of biodiversity and ecosystem services; third, the effects in return has brought about physical changes such as melting glacier, shore retreat, salinization, desertification, extreme events, melting ice sheets, dieback of forests and drying up of streams; fourth, those physical changes has imposed human impacts including reduction in crop yield and enhancing hunger, human disease, income loss in agriculture, fisheries and tourism, scarcity of water both in quantity and quality, voluntary and involuntary displacement, risk of instability and armed conflicts" [2, 17].

This chapter is framed on the following points:

1. Is agriculture, particularly unsustainable one, the cause of climate change? If so, what looks like its historical and projected contributions to climate change?
2. What are the return impacts of climate change on agriculture?
3. If agriculture and climate change have a reciprocal effect to each other, then what was their combined effects on global ecosystems?
4. If the global ecosystems are affected, then how the human wellbeing is affected altogether?
5. If the human wellbeing is affected, then what measure and actions should the global community would take to curb these problems?

2. Contribution of agriculture to greenhouse gases

2.1. Historical contribution

Agriculture is one of the contributors of greenhouse gases to climate change because agricultural activities are responsible for large-scale emissions of GHGs. Agriculture contributes to climate change by anthropogenic emissions of greenhouse gases and by the conversion of non-agricultural land such as forests to agricultural land [2].

The emission of GHGs from anthropogenic activities such as industrial process, land-use change and agriculture are the main drivers of climate change [2]. Agriculture's contribution to this was huge which took 14% of CO_2, 47% of CH_4 and 84% of N_2O of the global share of GHGs emission [18]. These are the most potent GHGs that are emitted from unsustainable agricultural practices. As compared to fossil fuels, the effect of land-use conversion on rising surface temperatures is an underestimated component of global warming [19]. Nonetheless, agriculture through tropical land use alone, mainly deforestation, contributed some 25% of CO_2 [8] from the total agriculture, forestry and other land-use (AFOLU) emissions. Fertilizer use in agriculture is another main human-made source of N_2O [8, 20]. The IPCC [15] definition of agriculture included cropland management, grazing land management/pasture improvement, management of agricultural organic soils, restoration of degraded lands, livestock management, manure/biosolid management and bioenergy production. These practices can result in the emissions of GHGs which in turn impacting agricultural development by contributing to climate change by the emissions of CH_4 from enteric fermentation and rice production, N_2O from soils, N_2O and CH_4 from manure management and biomass burning, and CO_2 emissions and removals in agricultural soils. The GHGs allow the penetration of incoming solar radiation but absorb the outgoing long-wave radiation from the Earth's surface and reradiate the absorbed radiation back to the surface of the Earth and by doing so they have caused global warming and climate change [2, 15].

Agricultural anthropogenic activities have increased and will continue to increase the concentration of GHGs in the atmosphere. As shown in **Table 1**, decadal average agriculture emissions grew, from 4.6 to 5.1 (4.8 ± 0.3) Gt CO_2 e year^{-1} in the 1990s to 5.0–5.5 (5.1 ± 0.3) Gt CO_2e year^{-1} in the 2000s, reaching 5.4 ± 0.3 Gt CO_2e year^{-1} in 2010 [20, 21].

2.2. Projected contribution

Agriculture is the main contributor of non-CO_2 GHGs such as CH_4 and N_2O which have a greater global warming potential than CO_2 (**Figure 1**). The EPA [21] studies showed that non-CO_2 emission from agriculture has increased from observed trend and continues to increase to their projections, that is, from observed emission of 5621.8 Mt. CO_2e in 1990 to projected emission of 6945 Mt. CO_2e in 2030 for the global estimate (**Table 2**). As the case in point for Ethiopia for example, the trend is an increase from 62.7 Mt. CO_2e to 133.9 Mt. CO_2e for the same period. Another report for Ethiopia's agricultural emission [22, 23] showed similar situation of increase from observed 127 Mt. CO_2e in 2010 to projected 275 Mt. CO_2 in 2030.

As outlined by Hristov et al. [24], livestock emissions took the lion's share of global non-CO_2 emissions through its land use and land-use change 2500 Mt. CO_2e, manure management 2200 Mt. CO_2e, animal production 1900 Mt. CO_2e, feed production (excluding carbon released from soil) 400 Mt. CO_2e and processing and international transport 30 Mt. CO_2e. Over the period 2001–2011, annual global emissions from enteric fermentation have increased by 11%, from 1858 Mt. CO_2e to 2071 Mt. CO_2e. These are projected to increase by 19% and 32% in 2030 and 2050, respectively, reaching more than 2500 Mt. CO_2e in 2080. Over the same period, annual emissions from manure management have increased about 10%, from 329 Mt. CO_2e

Year	Agriculture's non-CO_2 emissions ($MtCO_2e$)	Total non-CO_2 emissions ($MtCO_2e$)	Agriculture (%)
1990	5621.8	9771.2	57.5
1995	5501.8	9668.7	56.9
2000	5423.8	9896.5	54.8
2005	5798.5	10,780.7	53.8
2010	5998.8	11,387.3	52.7
2015	6271.2	12,166.0	51.5

Note: the calculation includes all non-CO_2 sources from energy, industrial process, agriculture and waste with few exceptions of CH_4 from hydroelectric reservoirs and abandoned coal mines, N_2O from industrial wastewater and F-GHG emissions from the manufacture of electrical equipment.

Table 1. Observed global total non-CO_2 emissions from all sources and from the agriculture sector [21].

http://dx.doi.org/10.5772/intechopen.79118

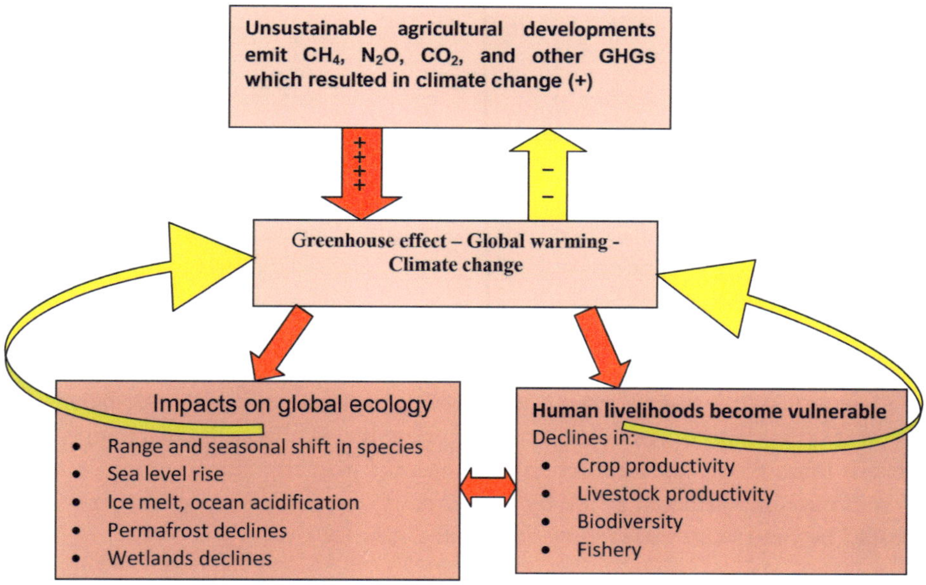

Figure 1. Reciprocate effects of agriculture and climate change on each other and the consequent impacts on ecology and human wellbeing (the red arrows show a direct and/or indirect impact (cause) on the other by which the one with positive sign showing causes for climate change; the vertical yellow arrow with negative sign shows the negative return impact of climate change on agriculture and the curved yellow arrows show that those decline in ecosystems and human wellbeing also have their own impacts on the climate system either directly or indirectly-developed based on IPCC [2, 15].

Year	Agriculture's non-CO_2 emissions ($MtCO_2e$)	Total non-CO_2 emissions ($MtCO_2e$)	Agriculture (%)
2020	6484.8	13,121.9	49.4
2025	6709.5	14,269.4	47.0
2030	6945.0	15,433.8	45.0

Note: for what the calculation includes see note under **Table 1**.

Table 2. Projected global total non-CO2 emissions from all sources and from the agriculture sector [21].

to 362 Mt. CO_2e and are projected to increase by 6% and 47% in 2030 and 2050, respectively, reaching more than 452 Mt. CO_2e in 2080 [8].

Several global studies suggested that at least until 2050 land-use change for crop production and livestock husbandry will be the dominant driver of terrestrial biodiversity loss in human-dominated regions [25–30]. Conversely, climate change is likely to dominate where human interventions are limited, such as in the tundra, boreal, cool conifer forests, deserts and savanna biomes. The effects of land-use change, particularly because of agriculture, on species through landscape fragmentation at the regional scale may further exacerbate impacts from climate change [2, 15].

3. Impacts of climate change on agriculture

Long-term fluctuations in weather patterns could have extreme impacts on agricultural production, slashing crop yields and forcing farmers to adopt new agricultural practices in response to altered conditions [31, 32]. As emphasized by Melillo et al. [33], some effects of climate change on agriculture include: loss of biodiversity in fragile environments/tropical forests, increased frequency of weather extremes (storms/floods/droughts), loss of fertile soil in coastal lands caused by rising sea levels, longer growing seasons in cool areas, more unpredictable farming conditions in tropical areas, and increase in incidence of pests and vector-borne diseases in livestock and dramatic changes in distribution and quantities of fish and sea foods. Climate is the primary determinant of agricultural productivity by which climate change is expected to influence crop and livestock production, hydrologic balances, input supplies and other components of agricultural systems [34, 35]. Climate change has posed a significant impact on crop production and livestock rearing. Studies showed that global wheat production is estimated to fall by 6% for each degree Celsius of further temperature increase and become more variable over space and time [36].

Agriculture is a victim of climate change (**Table 3**) because it is estimated that higher temperatures could reduce crop yield by 10–20% in sub-Saharan Africa by 2050. In return, agricultural development is one of the causes of climate change because it is responsible for 10–12% of human-generated GHGs emissions each year and much more (30%) if humans take into account the clearance of forests to make way for crops and livestock [37, 38]. Specific climate-related impacts also have national or regional level impacts on agriculture. Bearing in mind that one of the indictors of global climate change is an increase in global temperature and one of the first requirements for hurricanes (among other things such as pressure difference in the wind current) to be crated is warming of the ocean water more than 80°F, and it is possible to correlate that climate change aggravates hurricanes to happen frequently [39–41]. For example, Hurricane Mitch which hit Central American countries such as Honduras, Nicaragua, Belize, Costa Rica, El Salvador, Guatemala and Panama has brought the most severe damage on the export and subsistence agricultural sectors with an estimated 70% of total damage or US$1.7 billion. It has brought in 58% corn lost, 24% sorghum, 14% rice, 6% beans, 85% bananas, 60% sugarcane, 28% African Palm and 18% coffee [6, 19, 42, 43].

A study by IFPRI [44] showed climate change is supposed to have reduction in net crop revenue by (−28% to −79%), (−7% to −32%), (−12% to −17%), (−11% to −12%) and (−4% to −7%) in Central Africa, West Africa, southern Africa, East Africa and North Africa, respectively. In Ethiopia, the study by Deressa [45] showed that a unit increase in temperature during summer and winter would reduce net revenue per hectare by US$177.62 and 464.71, respectively, whereas the marginal impact of increasing precipitation during spring would increase net revenue per hectare by US$225.09. In another similar case for example, the 2008–2011 droughts in Kenya caused a total of USD 10.7 billion in damages and losses in agriculture sector and subsectors [46]. As Brown et al. [4] dictated, by being affecting agricultural production and productivity, climate change is very likely to affect global, regional and local food security by disrupting food availability, decreasing access to food and making food utilization more difficult.

Regions	Impacts of climate change on agriculture
Asia and Pacific	• Freshwater availability in Central, South, East and Southeast Asia is likely to decrease. • Temperature increases will lead to a substantial increase in demand for irrigation water for sustained productivity in arid, semiarid Asia and South and East Asia. • Land suitable for crop cultivation is expected to increase in East and Central Asia, but decrease in other areas, especially in South Asia. • Crop yields could increase in East and Southeast Asia, while they could decrease in Central and South Asia even considering the fertilization effects of CO_2. • There will likely be a northward shift of agricultural zones. • Heat stress and limited pasture availability would limit the expansion of livestock numbers.
Europe and Central Asia	• Countries in the more temperate and polar regions are likely to benefit. • Countries in midlatitudes will benefit at first but will begin to be affected negatively if temperatures rise by more than 2.5°C. • The combination of temperature increase and increasing CO_2 concentration will result in slightly positive agricultural development in southeastern Europe, while the Mediterranean area and southwest Balkans will suffer. • Central Asia, dependent on irrigation and with high interannual variations in yields, can be affected by climate extremes and decrease in water availability. • Cattle and small livestock could suffer from increasing heat stress and spread of diseases.
Near East	• Maize yields in North Africa would suffer first with rising temperatures, followed by Western Asia and the Middle East. • Water availability would decrease in most of the region, although it may slightly increase in some areas, such as most of Sudan, Somalia and southern Egypt. • Temperature increase may lead to increased pasture production in midlatitudes, with increases in livestock production. • Warmer winters may benefit livestock, while greater summer heat stress can have negative effects.
Africa	• The number of extremely dry and wet years is expected to increase in sub-Saharan Africa during this century. • Drying is expected in the Mediterranean area and in much of southern Africa. • Rainfall may increase in east and west Africa. • Some areas, such as the Ethiopian highlands, could benefit from a longer growing season. • Rangeland degradation and more frequent droughts may lead to reduced forage productivity and quality, particularly in the Sahel and southern Africa.
Latin America and Caribbean	• In temperate zones, such as southeastern South America, yield of certain crops such as soybean and wheat will increase. • As a result of increased thermal stress and drier soils, productivity in tropical and subtropical regions is expected to decline. • In arid zones, such as central and northern Chile and northeastern Brazil, the salinization and desertification of agricultural land will possibly increase. • Rain fed agriculture in semiarid zones will face increasing risks of losing crops. • In temperate areas, pasture productivity may increase benefiting livestock production.

Source: FAO [47].

Table 3. Selected possible regionalized impacts of climate change on agriculture.

4. Impacts of climate change and agriculture on ecosystems

As already indicated in Section 2 of this chapter, agriculture is one of the contributors of greenhouse gases to climate change so does has a contribution to any impact of climate change on global ecosystem. It also affects global ecology through its land-use changes particularly in tropical forest ecosystem change by deforestation for agriculture [28]. Several studies [31, 48–53] showed that the impacts of climate change on global ecosystems are apparent, and future change is likely to be dramatic. By the mid of the twenty-first century, scientific evidence indicated the likelihood of global temperature rising between 3 and 4°C above the

ΔT	Impacts on terrestrial and aquatic ecosystems	References
1.6	Bioclimatic envelopes eventually exceeded leading to: • transformation of 10% of global ecosystems; • loss of 47% wooded tundra, 23% cool conifer forest, 21% scrubland, 15% grassland/steppe, 14% savanna, 13% tundra and 12% temperate deciduous forest; and • ecosystems variously lose 2–47% areal extent.	[54]
2.4	63 of 165 rivers studied lose more than 10% of their fish species	[57]
2.5	Sink service of terrestrial biosphere saturates and begins turning into a net carbon source	[58, 59]
2.7	Bioclimatic envelopes exceeded leading to: • eventual transformation of 16% of global ecosystems; • loss of 58% wooded tundra, 31% cool conifer forest, 25% scrubland, 20% grassland/steppe, 21% tundra, 21% temperate deciduous forest and 19% savanna; and • ecosystems variously lose 5–66% of their areal extent.	[54]
3.0	66 of 165 rivers studied lose more than 10% of their fish species	[57]
3.1	Extinction of remaining coral reef ecosystems (overgrown by algae)	[60]
3.3	Reduced growth in warm-water aragonitic corals by 20–60% and 5% decrease in global phytoplankton productivity	[60–62]
3.4	6–22% loss of coastal wetlands, large loss of migratory bird habitat particularly in the USA, Baltic and Mediterranean	[63, 64]
3.5	Predicted extinction of 15–40% endemic species in global biodiversity hotspots (case "narrow biome specificity")	[65]
3.7	Bioclimatic envelopes exceeded leading to: • eventual transformation of 22% of global ecosystems; • loss of 68% wooded tundra, 44% cool conifer forest, 34% scrubland, 28% grassland/steppe, 27% savanna, 38% tundra and 26% temperate deciduous forest; and • ecosystems variously lose 7–74% areal extent.	[54]

Source: IPCC [2, 15].

Table 4. Projected impacts of climate change on global ecosystems as reported in the literature for different levels of global mean annual temperature rise, ΔT (°C), relative to preindustrial (PI) climate.

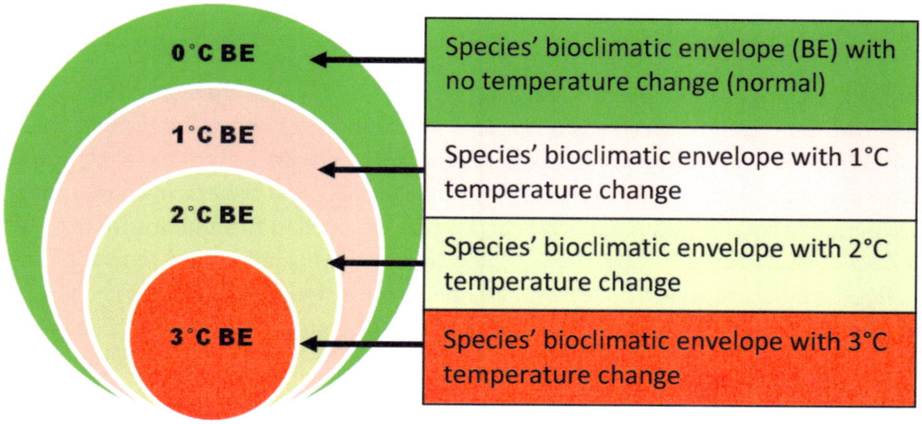

Figure 2. Schematic representation of change in bioclimatic envelope of a species with respect to climate change developed based on Malcolm and Pitelka [56].

preindustrial level [2]. As described in **Table 4**, the projected impact will lead to serious consequences for humans and ecosystems due to dangerous sea level rise, unprecedented heat waves, severe drought and major floods in many parts of the world [15].

If global mean temperature (GMT) change is more than 3°C, very few ecosystems can adapt while most of regional and global ecosystems will be at risk [54]. Climate change could have a profound impact on biodiversity directly through changes in temperature and precipitation and indirectly in the ways it might affect land use and nutrient cycles, ocean acidification and the prospects for invasion of alien species into new habitats [9]. Climate change leads narrow bioclimatic envelope (**Figure 2**)—the range of climatic conditions within which a species can survive and grow [55]. In other words, when the global mean temperature change increases beyond 3.5°C, most of the species have very few suitable area for their survival and will become extinct [54].

Climate change is happening on a global scale, but the ecological impacts are often local and vary from place to place [2, 15]. These impacts can include expansion of species into new areas, intermingling of formerly nonoverlapping species and even species extinctions. Two important types of ecological impacts of climate change have been observed. First, shifts in species' ranges (the locations in which they can survive and reproduce) and second, shifts in phenology (the timing of biological activities that take place seasonally). Other ecological impacts of climate change include changes in growth rates, in the relative abundance of species, in processes like water and nutrient cycling, and in the risk of disturbance from fire and invasive species. If a level of global warming occurs in the range from 3.6 to 5.4°F—somewhere in the low-to-mid projected range—it is estimated that about 20–30% of studied species could risk extinction in the next hundred years. Given that there are approximately 1.7 million identified species on the globe, this ratio would suggest that some 3–6 hundred thousand species could be committed to extinction [13].

5. Impacts of climate change on human livelihoods

Several hundred million people are seriously affected by climate change today, with several hundred thousand annual deaths [26, 46, 66]. Some human impacts of climate change [15] includes: hundreds of millions of people exposed to increased water stress; complex, localized negative impacts on small holders, subsistence farmers and fishers; millions more people could experience coastal flooding each year; increasing burden from malnutrition, diarrheal, cardio-respiratory and infectious diseases; and increased morbidity and mortality from heat waves, floods and droughts. The World Health Organization's [67] global burden of disease study showed that long-term consequences of climate change affected over 325 million people in 2004. By the year 2030, the lives of 660 million people are expected to be seriously affected (increase of 103%) either by natural disasters caused by climate change or through gradual environmental degradation (**Figure 3**). In addition to what has been described in **Table 5**, human impacts of climate change include scarcity of freshwater resources, weather-related disasters, food insecurity due to agricultural loss, migration and displacement due to loss of settlements which can be exemplified by the following extreme events.

- Flooding in Pakistan severely affected crops and livestock, where the crops were either partially or completely submerged and the livestock suffered from a lack of fodder availability. A total country wide loss of US$1840 million was expected to have occurred in the agricultural sector [68].
- Flooding and drought combined in Mozambique adversely affected the livelihood of the rural farmers. In the year 2007 alone, Mozambique experienced a total economic loss and damage of $71,000 from severe flooding. Crop cultivation, livestock rearing and fishing were the most prominent sources of income for rural livelihood and are the most affected by climate-related risks [69].
- The main sources of livelihood in the flood prone regions of Kenya are crop cultivation, livestock rearing and other non-agricultural activities such as fishing, small-scale trade and manual labor. Flooding and drought in these low-lying areas had increased severely and caused approximate monetary loss of about US$0.5 billion per year which is equivalent to 2% of the country's GDP. This cost is expected to rise and eventually claim 3% of Kenya's GDP by 2030 [70].
- In severely drought prone regions of Gambia, the varying level of rainfall, shorter duration of the rainy season along with rising temperatures had resulted in severe calamity for its community that was mostly reliant on agriculture for their livelihoods. The residual damages from climate change in Gambia ranged between US$123 million and US$130 million per year in the near term and estimated to range from US$955 million to US$1.0 billion for the period 2070–2099 [71].
- Ninety percent of Burkina Faso's population is engaged in agriculture and livestock sectors which are very sensitive to climate variability. The range of average crop production loss due to drought was reported to be between US$577 and US$636 per household, whereas the range of average livestock loss was found to be between US$1922 and 8759 per herder in the region [69].

http://dx.doi.org/10.5772/intechopen.79118

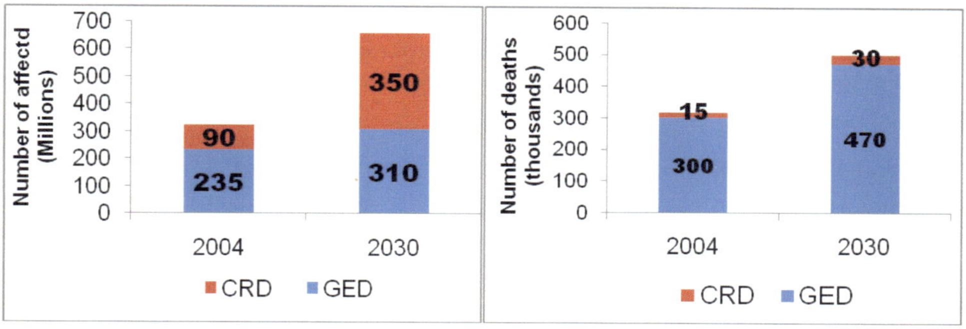

Figure 3. The impact of climate change due to gradual environmental degradation (GED) and climate-related disasters (CRD)-[67].

Year	Climate change-related disaster	Country	People killed	Economic loss
2007	Cyclone Sidr	Bangladesh	3400	\$1.6 Billion
2008	Cyclone Nargis	Myanmar	150,000	\$4.0 Billion
2005	Hurricane Katrina	USA	1800	\$100 Billion
2009	Cyclone Aila	Bangladesh	190	\$170 Million
2013	Typhoon Haiyan	Philippines	7986	\$10 Billion
1973	Drought	Ethiopia	100,000	\$76 Million
2011	Cyclone Yasi	Australia	1	\$3.6 Billion
1998	Hurricane Mitch	Honduras, Nicaragua, Belize, Costa Rica, El Salvador, Guatemala, Panama	11,000	\$ 5 Billion
2005	Hurricane Wilma	Mexico, USA, Bahamas, Cuba, Haiti, Jamaica	63	\$ 29.4 Billion

Table 5. Few examples of the impacts of climate change on human livelihoods [17, 66].

6. Recommendations for action

- Basing the IPCC reports as the fundamental principle, "countries should act now, act together and act differently - do better actions beyond the 'paper' - on the stabilization of greenhouse gases concentrations in the atmosphere at a level that would prevent dangerous anthropogenic interference with the climate system within a time frame sufficient to allow ecosystems to adapt naturally to climate change," to ensure that food production is not threatened and to enable economic development to proceed in a sustainable manner so as to calm down the impacts of climate change;
- Countries shall promote sustainable forms of agriculture in light of climate change in order to promote sustainable agricultural development which sustainably increases productivity and resilience, reduces/removes greenhouse gases emissions and enhances achievement of national food security and development goals in order to minimize the effects of agriculture to climate change and vice versa;

- Climate change abatement requires environmental conservation and global partnership that related to two of the Millennium Development Goals (MDGs): ensure environmental sustainability and develop a global partnership for development which are reconciled with the Sustainable Development Goals (SDGS);
- To meet developmental success by overcoming the challenges of climate change to agriculture, it requires a comprehensive approach of technical, institutional and financial innovations, so that both adaptation and mitigation strategies are consistent with efforts to safeguard food security, maintain ecosystem services, provide carbon sequestration and reduce emission in agricultural landscapes;
- Productive and ecologically sustainable agriculture with strongly reduced greenhouse gases emissions is fundamental so as to reduce trade-offs among agricultural development to fulfill food security, climate change and ecosystem degradation;
- Reports and studies of observational evidence from all continents and most oceans showed that many natural systems are being affected by regional climate changes, particularly temperature increases. To curb these climate change impacts, there is a need of human solutions for human causes: the world should invest in minimizing the amount of climate change that occurs and in adapting to the changes that cannot be avoided.

Author details

Zenebe Mekonnen

Address all correspondence to: zenebemg2014@gmail.com

Wondo Genet College of Forestry and Natural Resources, Hawassa University, Shashemene, Ethiopia

References

[1] OECD (Organization for Economic Co-operation and Development). The Benefits of Climate Change Policies. Paris, France: OECD; 2004

[2] IPCC. Climate Change 2013: The Physical Science Basis. In: Stocker TF, Qin D, Plattner GK, Tignor M, Allen SK, Boschung J, Nauels A, Xia Y, Bex V, Midgley PM, editors. Contribution of Working Group I to the Fifth Assessment Report of the Intergovernmental Panel on Climate Change. Cambridge, UK and New York, USA: Cambridge University Press; 2013. 1535 pp

[3] FAO. The State of Food and Agriculture—Women in Agriculture: Closing the Gender Gap for Development. Italy: Rome; 2011

[4] Brown ME, Antle JM, Backlund P, Carr ER, Easterling WE, Walsh MK, Ammann C, Attavanich W, Barrett CB, Bellemare MF, Dancheck V, Funk C, Grace K, Ingram JSI,

Jiang H, Maletta H, Mata T, Murray A, Ngugi M, Ojima D, O'Neill B, Tebaldi C. Climate Change, Global Food Security, and the U.S. Food System. 2015. DOI: 10.7930/J0862DC7

[5] IPCC. Climate change 2001: The scientific basis. In: Houghton JT, Ding Y, Griggs DJ, Noguer M, van der Linden PJ, Dai X, Maskell K, Johnson CA, editors. A Report of the Working Group I of the Intergovernmental Panel on Climate Change. Cambridge, UK: Cambridge University Press; 2001

[6] ECLAC (United Nations Economic Commission for Latin America and the Caribbean). Latin America and the Caribbean: Natural Disaster Impacts on Development, 1972-1999; 1999

[7] CGIAR (the Consultative Group on International Agricultural Research). Agriculture and Rural Development Day 2012: Lessons in Sustainable Landscapes and Livelihoods; 2012

[8] FAO. Agriculture, Forestry and Other Land Use Emissions by Sources and Removals by Sinks: 1990-2011 Analysis. Rome, Italy: FAO; 2014

[9] Bennetzen EH, Smith P, Porter JR. Agricultural production and greenhouse gas emissions from world regions—The major trends over 40 years. Global Environmental Change. 2016;**37**:43-55

[10] Elasha BO, Elhassan NG, Ahmed H, Zakieldin S. Sustainable livelihood approach for assessing community resilience to climate change: Case studies from Sudan. AIACC Working Paper No. 17. Khartoum, Sudan: Higher Council for Environment and Natural Resources; 2005

[11] Tol RSG. The economic effects of climate change. Journal of Economic Perspectives. 2009;**23**(2):29-51

[12] CIDSE. Agriculture: From Problem to Solution-Achieving the Right to Food in a Climate-Constrained World. Brussels, Belgium: CIDSE; 2012

[13] NAS. Climate Intervention: Carbon Dioxide Removal and Reliable Sequestration. Washington, DC, USA: National Research Council of the National Academies, The National Academies Press; 2015

[14] Bellard C, Leclerc C, Courchamp F. Impact of sea level rise on the 10 insular biodiversity hotspots. Global Ecology and Biogeography. 2014;**23**:203-212

[15] IPCC. Climate change 2007: Impacts, adaptation and vulnerability. In: Parry ML, Canziani OF, Palutikof JP, van der Linden PJ, Hanson CE, editors. Contribution of Working Group II to the Fourth Assessment Report of the Intergovernmental Panel on Climate Change. Cambridge, UK: Cambridge University Press; 2007. 976 pp

[16] Nye J. Climate Change and its Effects on Ecosystems, Habitats and Biota State of the Gulf of Maine Report; 2010

[17] GHF. Climate change: The anatomy of a silent crisis. Human Impact Report. Geneva, Switzerland: Global Humanitarian Forum; 2009

[18] Smith P, Martino D, Cai Z, Gwary D, Janzen H, Kumar P, McCarl B, Ogle S, O'Mara F, Rice C, Scholes B, Sirotenko O, Howden M, McAllister T, Pan G, Romanenkov V, Schneider U, Towprayoon S. Policy and technological constraints to implementation of greenhouse gas mitigation options in agriculture. Agriculture, Ecosystems and Environment. 2007;**118**:6-28

[19] Stonich S. I am Destroying the Land: The Political Ecology of Poverty and Environmental Destruction in Honduras. Boulder, Colorado, USA: Westwiew Press; 1993

[20] Tubiello FN, Salvatore M, Ferrara AF, House J, Federici S, Rossi S, Biancalani R, Golec C, Jacobs H, Flammini A, Prosperi P, Cardenas-Galindo P, Schmidhuber J, Sanchez MJS, Srivastava N, Smith P. The contribution of agriculture, forestry and other land use activities to global warming, 1990-2012. Global Change Biology. 2015;**21**(7):1655-1660. DOI: 10.1111/gcb.12865

[21] EPA. Global Anthropogenic Non-CO_2 Greenhouse Gas Emissions: 1990-2030. Washington, DC, USA: EPA; 2012

[22] DEA (Danish Energy Agency), The Organization for Economic Co-Operation and Development (OECD), The UNEP Risø Centre (URC). National Greenhouse Gas Emissions Baseline Scenarios: Learning from Experiences in Developing Countries; 2013

[23] FDRE. The Path to Sustainable Development: Ethiopia's Climate-Resilient Green Economy Strategy. Addis Ababa, Ethiopia; 2011

[24] Hristov AN, Oh J, Lee C, Meinen R, Montes F, Ott T, Firkins J, Rotz A, Dell C, Adesogan A, Yang W, Tricarico J, Kebreab E, Waghorn G, Dijkstra J, Oosting S. Mitigation of greenhouse gas emissions in livestock production—A review of technical options for non-CO_2 emissions. In: Gerber PJ, Henderson B, Makkar HPS, editors. FAO Animal Production and Health Paper No. 177. Rome, Italy: FAO; 2013

[25] Sala OE, Chapin IFS, Armesto JJ, Berlow E, Bloomfield J, Dirzo R, Huber Sanwald E, Huenneke LF, Jackson RB, Kinzig A, Leemans R, Lodge DH, Mooney HA, Oesterheld M, Leroy Poff N, Sykes MT, Walker BH, Walker M, Wall DH. Global biodiversity scenarios for the year 2100. Science. 2000;**287**(5459):1770-1774

[26] UNEP (United Nations Environment Programme). Global Environment Outlook 3: Past, Present and Future Perspectives. London, UK: Earthscan Publications Ltd.; 2002. pp. 426+Xxx

[27] Gaston KJ, Blackburn TM, Klein GK. Habitat conversion and global avian biodiversity loss. Proceedings of the Royal Society, London B. 2003;**270**:1293-1300

[28] Jenkins M. Prospects for biodiversity. Science. 2003;**302**:1175-1177

[29] Scharlemann JPW, Green RE, Balmford A. Land-use trends in endemic bird areas: Global expansion of agriculture in areas of high conservation value. Global Change Biology. 2004;**10**:2046-2051

[30] Sala OE, Detlef van Vuuren D, Pereira HM, Lodge D, Alder J, Cumming G, Dobson A, Wolters V, Xenopoulos MA. Biodiversity across scenarios. In: Ecosystems and Human Well-being, Vol. 2: Scenarios. Washington, DC, USA: Island Press; 2005. pp. 375-408

[31] Megersa B, Markemann A, Angassa A, Ogutu JO, Piepho H, Zaráte AV. Impacts of climate change and variability on cattle production in southern Ethiopia: Perceptions and empirical evidence. Agricultural Systems. 2014;**130**:23-34

[32] Chikodzi D. Crop yield sensitivity to climatic variability as the basis for creating climate resilient agriculture. American Journal of Climate Change. 2016;**5**:69-76

[33] Melillo JM, Terese TCR, Gary WY, editors. Highlights of Climate Change Impacts in the United States: The Third National Climate Assessment. U.S. Global Change Research Program; Washington, DC, USA. 2014. pp. 148

[34] Adams RM, Hurd BH, Lenhart S, Leary N. Effects of global climate change on agriculture: An interpretative review. Climate Research. 1998;**11**:19-30

[35] Aydinalp C, Cresser MS. The effects of global climate change on agriculture. American-Eurasian Journal of Agricultural & Environmental Sciences. 2008;**3**(5):672-676

[36] Asseng S, Ewert F, Martre P, Rötter RP, Lobell DB, Cammarano D, Kimball BA, Ottman MJ, Wall GW, White JW, Reynolds MP, Alderman PD, Prasad PV, Aggarwal PK, Anothai J, Basso B, Biernath C, Challinor AJ, De Sanctis G, Doltra J, Fereres E, Garcia-Vila M, Gayler S, Hoogenboom G, Hunt LA, Izaurralde RC, Jabloun M, Jones CD, Kersebaum KC, Koehler AK, Müller C, Kumar SN, Nendel C, O'Leary G, Olesen JE, Palosuo T, Priesack E, Rezaei EE, Ruane AC, Semenov MA, Shcherbak I, Stöckle C, Stratonovitch P, Streck T, Supit I, Tao F, Thorburn PJ, Waha K, Wang E, Wallach D, Wolf J, Zhao Z, Zhu Y. Rising temperatures reduce global wheat production. Nature Climate Change. 2015;**5**:143-147

[37] Meridian Institute. Agriculture and climate change policy brief: Main issues for the UNFCCC and beyond. In: Lee D, editor. Adapted from "Agriculture and Climate Change: A Scoping Report" by Bruce Campbell, Wendy Mann, Ricardo Meléndez-Ortiz, Charlotte Streck, Timm Tennigkeit, and Sonja Vermeulen. 2011. Available at: www.climate-agriculture.org

[38] Pye-Smith C. Promoting climate-smart agriculture in ACP countries. CTA Policy Brief No. 9. Wageningen, The Netherlands; 2012

[39] Willoughby HE. Hurricane heat engines. Nature. 1999;**401**:649-650

[40] Knutson TR, McBride JL, Chan J, Emanuel K, Holland G, Landsea C, Held I, Kossin JP, Srivastava AK, Sugi M. Tropical cyclones and climate change. Nature Geoscience. 2010; **3**:157-163

[41] Kang N, Elsner JB. Trade-off between intensity and frequency of global tropical cyclones. Nature Climate Change Letters. 2015;**5**:661-664. DOI: 10.1038/NCLIMATE2646

[42] Charvériat C. Natural disasters in Latin America and the Caribbean: An overview of risk. Working Paper 434. New York, USA: Inter-American Development Bank; 2000

[43] UNEP. Global Environment Outlook: Latin America and the Caribbean, GEO LAC 3. Panama City, Panama: UNEP; 2010

[44] IFPRI (International Food Policy Research Institute). Micro-level practices to adapt to climate change for african small-scale farmers. IFPRI Discussion Paper 00953. Addis Ababa, Ethiopia; 2010

[45] Deressa TT. Measuring the Economic Impact of Climate Change on Ethiopian Agriculture: Ricardian Approach. The World Bank Development Research Group, Sustainable Rural and Urban Development Team. Policy Research Working Paper 4342. Addis Ababa, Ethiopia; 2007

[46] FAO. The Impact of Natural Hazards and Disasters on Agriculture and Food Security and Nutrition: A Call for Action to Build Resilient Livelihoods. Rome, Italy: FAO; 2015

[47] FAO. FAO-Adapt Framework Programme on Climate Change Adaptation. Rome, Italy; 2011

[48] Norris K. Agriculture and biodiversity conservation: Opportunity knocks. Conservation Letters. 2008;**1**:2-11

[49] Williams AP, Allen CD, Millar CI. Forest responses to increasing aridity and warmth in the southwestern United States. Proceedings of the National Academy of Sciences. 2010;**107**:21289-21294

[50] Comer PJ, Young B, Schulz K, Kittel G, Unnasch B, Braun D, Hammerson G, Smart L, Hamilton H, Auer S, Smyth R, Hak J. Climate change vulnerability and adaptation strategies for natural communities: Piloting methods in the Mojave and Sonoran deserts. In: Report to the U.S. Fish and Wildlife Service. Arlington, VA: NatureServe; 2012

[51] Notaro M, Mauss A, Williams JW. Projected vegetation changes for the American southwest: Combined dynamic modeling and bioclimatic envelope approach. Ecological Applications. 2012;**22**:1365-1388

[52] Friggens MM, Bagne KE, Finch DM, Falk D, Triepke J, Lynch A. Review and recommendations for climate change vulnerability assessment approaches with examples from the Southwest. General Technical Report. RMRS-GTR-309. Fort Collins, CO: U.S. Department of Agriculture, Forest Service, Rocky Mountain Research Station; 2013. 106 p

[53] Luedeling E, Muthuri C, Kindt R. Ecosystem vulnerability to climate change: A literature review. Working Paper 162. Nairobi, Kenya: World Agroforestry Centre; 2013. DOI: 10.5716/WP13034.PDF

[54] Leemans R, Eickhout B. Another reason for concern: Regional and global impacts on ecosystems for different levels of climate change. Global Environmental Change. 2004;**14**: 219-228

[55] Pearson RG, Dawson TP. Predicting the impacts of climate change on the distribution of species: Are bioclimate envelope models useful? Global Ecology & Biogeography. 2003;**12**:361-371

[56] Malcolm JR, Pitelka LF. Ecosystems and Global Climate Change: A Review of Potential Impacts on U.S. Terrestrial Ecosystems and Biodiversity Prepared for the Pew Center on Global Climate Change. Washington, DC, USA; 2000

[57] Xenopoulos MA, Lodge DM, Alcamo J, Marker M, Schulze K, Van Vuuren DP. Scenarios of freshwater fish extinctions from climate change and water withdrawal. Global Change Biology. 2005;**11**:1557-1564

[58] Lucht W, Schaphoff S, Erbrecht T, Heyder U, Cramer W. Terrestrial vegetation redistribution and carbon balance under climate change. Carbon Balance Management. BioMed Central Ltd. 2006;**1**(6). DOI: 10.1186/1750-0680-1-6

[59] Schaphoff S, Lucht W, Gerten D, Sitch S, Cramer W, Prentice IC. Terrestrial biosphere carbon storage under alternative climate projections. Climatic Change. 2006;**74**:97-122

[60] Hoegh-Guldberg O. Climate change, coral bleaching and the future of the world's coral reefs. Marine and Freshwater Research. 1999;**50**:839-866

[61] Raven J, Caldeira K, Elderfield H, Hoegh-Guldberg O, Liss P, Riebesell U, Shepherd J, Turley C, Watson A. Ocean Acidification Due to Increasing Atmospheric Carbon Dioxide, Royal Society Policy Document. London, UK: The Royal Society; 2005

[62] Cox PM, Betts RA, Jones CD, Spall SA, Totterdell IJ. Acceleration of global warming due to carbon-cycle feedbacks in a coupled climate model. Nature. 2000;**408**:184-187

[63] Nicholls RJ, Hoozemans FMJ, Marchand M. Increasing flood risk and wetland losses due to global sea-level rise: Regional and global analyses. Global Environmental Change. 1999;**9**:S69-S87

[64] Najjar RG, Walker HA, Barron EJ, Bord RJ, Gibson JR, Kennedy VS, Knight CG, Megonigal JP, O'Connor RE, Polsky CD, Psuty NP, Richards BA, Sorenson LG, Steele EM, Swanson RS. The potential impacts of climate change on the mid-Atlantic coastal region. Climate Research. 2000;**14**:219-233

[65] Malcolm JR, Liu C, Miller L, Allnut T, Hansen L, editors. Habitats at risk: Global warming and species loss in globally significant terrestrial ecosystems. WWF World Wide Fund for Nature, Gland; 2002. 40 pp

[66] Guha-Sapir D, Vos F, Below R, Ponserre S. Annual Disaster Statistical Review 2011: The Numbers and Trends. Brussels, Belgium: CRED; 2012

[67] WHO (World Health Organization). Global Burden of Disease 2004 Update. Geneva, Switzerland; 2004

[68] Hasnain Z. Pakistan: Flood Impact Assessment. Islamabad: International Federation of Red Cross and Red Crescent Societies; 2011

[69] Mathew LM, Akter S. Loss and damage associated with climate change impacts. 2014. Available from: http://irri.academia.edu/SoniaAkter/Working-Papers

[70] SEI (Stockholm Environmental Institute). Economics of Climate Change in Kenya. Oxford: The Stockholm Environment Institute; 2009

[71] Njie M, Gomez BE, Hellmuth ME, Callaway JM, Jallow BP, Droogers P. Making economic sense of adaptation in upland cereal production in the Gambia. In: Leary N, Conde C, Kulkarni J, Nyong A, Pulhin J, editors. Climate Change and Vulnerability and Adaptation. London: Earthscan Publications Ltd; 2007. pp. 131-146

Chapter 8

Energy Mining, Earth's Thermal Insulation Damaged and Trigger Climate Change

Yao Mu and Xinzhi Mu

Additional information is available at the end of the chapter

http://dx.doi.org/10.5772/intechopen.80537

Abstract

Fossil energy is the product of a series of complex chemical reactions inside the earth under high temperature and pressure. Where there is fossil energy, there must be a huge heat reservoir. The vast majority of coal, oil and gas are found in sedimentary basins with abundant geothermal resources. There is no "sea of oil" or "sea of gas" in the Earth's crust. Oil, natural gas, shale gas, etc. exist underground in rock pores, cracks, caves, faults, sand grains where like a huge "capillary network". Some cracks and faults reach deep into the entire crust. Oil, natural gas and shale gas seal off these pores, cracks, faults and sand layers, effectively preventing excessive leakage of heat from the ground. The enormous pressure of oil, gas and shale gas in the Earth's crust counteracts the thermal pressure in the Earth's interior, reaching a dynamic equilibrium. Once the oil, gas and shale gas is out of the ground, due to the loss of heat insulation and heat insulation material, the heat will eventually reach the surface from the Earth's interior, causing the Earth's crust "fever". A large number of water vapor, carbon dioxide, methane, etc. Greenhouse gas from the crust into the atmosphere and ocean, destroyed the energy balance of the atmosphere. This article aims to find out the real causes of climate change. By collecting materials from published academic documents, it is clarified that the man-made damage to the Earth's crust heat insulation seal is the truth of climate change. Therefore, the following conclusions are drawn: the thermal insulation of the Earth's crust is damaged by mining fossil energy (coal, oil, natural gas, shale gas, oil shale, gas hydrate, etc.), too much heat from the Earth's interior is pouring into the Earth's surface, causing the Earth's crust temperature and sea temperature to rise, trigger climate change and ecological disasters. Large amounts of water vapor have entered space, resulting rainfall and snow in some areas to exceed historical limits several times. Global soil and oceans degradation year by year.

Keywords: fossil energy extraction, terrestrial heat flow, Earth's crust, underlying surface, climate change

1. Real culprit in climate change today

At present, the ecological and geological disasters, climate change, extreme meteorological disasters, etc., and fossil energy, mining, crustal heat insulation layer destruction, the terrestrial heat flow increased, leading to the Earth's crust and ocean temperature caused climate change tend to be ignored.

Although the cause of climate change opinions vary, I have been committed to the research of climate change, finally come to conclusion is industrial mining fossil fuels (coal, oil, natural gas, oil shale, natural gas hydrate, etc.), destroyed the Earth's crust inside insulation, excessive heat from the interior of the earth to the surface, this is the main cause of climate change. Atmospheric events are becoming more extreme as CO_2, CH_4 and other gases stored in the Earth's crust are released into the atmosphere due to heat in the Earth's crust and an increase in the temperature of its underground surface. The Earth's crust and underlying surfaces are being heated in irregular ways! It is for this reason, not only the ecological environment rapidly deteriorating, and because of excessive heat into the atmosphere and ocean, increasingly frequent extreme weather events occur, constantly refresh historical extremum. Fossil energy is the product of a series of complex chemical reactions under the long-term high temperature and pressure in the earth. Therefore, where there is fossil energy, there are a large number of thermal resources. But does not exist in the crust "oil sea" or "gas sea", rock pores, fissures, karst cave, fault and thick sandstone exist in oil and natural gas fossil, formed a huge "capillary network". Although the fossil energy constitute only a small part of the Earth's surface, exploiting the maximum depth of only 5000 m, but the fossil energy and gas seam sealed effectively to rock pore, fracture and karst cave, fault and coal, prevent the excessive leakage of terrestrial heat flow. The enormous pressure of oil, gas and shale gas in the Earth's crust counteracts the heat pressure in the Earth's interior and achieves a dynamic equilibrium. Once the oil, gas and shale gas have been mined, the Earth's interior heat flow because of lost the thermal insulation layer, eventually to travel to the surface, causing the Earth's crust "fever", cause the ecological and geological disasters. A large amount of extremely dispersed heat forces gases from the Earth's crust, such as water vapor, CO_2 and CH_4, into the atmosphere and ocean, damaging the energy balance of the atmosphere and causing climate change and meteorological disasters. With the increase of ocean temperature, air humidity increases, the recorded the strongest typhoon, hurricanes and tropical cyclones, and recorded the strongest local rainfall, snow, drought, heat and cold. These are expected to become more frequent as the weather becomes more extreme and violent. Even if humans stop emitting greenhouse gases, global change will still continue for a long time because of this reason. The ground is a big tree, the underground the capillary roots has a football field yet. The same is true even though fossil fuels account for 1% of the total Earth's surface area, but exist in rock pores, cracks, karst caves, faults and grits through the crust, forming a huge "capillary network" far more than 1%.

The arctic is sparsely populated, artificial GHGs emissions are almost non-existent, the concentration of CO_2 in the atmosphere is very low. But the region is rich in oil and gas, and the massive mining of the surrounding countries (Russia, Alaska, Norway, Canada, Denmark, etc.) has accelerated the melting of arctic sea ice, which is a living example.

Although the oil/gas extracted rock pores are filled with water, the thermal insulation layer is destroyed, and the geothermal energy travels up in all directions. Moreover, the thermal insulation effect of oil and gas is far greater than that of water. Pandora's box once opened, will never be closed again.

2. Root cause of climate change is Earth's solid part

It is generally accepted in the meteorological community that climate change that occurs for longer than 3 months is not caused by factors within the atmosphere but outside of the atmosphere.

Tang et al. [1, 2] suggested that "The fundamental cause of climate change is the solid part of the earth," and atmospheric change is a responsive (or adaptive) state. The authors propose a "geocentric theory of climate change". Ground temperature is a measurement of the physical quantity of surface thermal energy. The change of ground temperature is more conservative than the change of temperature, which is caused by hysteresis. The change of ground temperature as part of the ground surface system, can cause the change of many physical quantities in the atmosphere and ocean, this shows that the solid circulation in various thermal processes are involved in the process of climate change [3]. Ground temperature and sea temperature a direct representation of surface heat. The heat source of the Earth's temperature ocean temperature are solar radiation and heat diffusion within the crust. According to the influence of heat source on the crust, the crust surface can be divided into variable temperature zone, stable temperature zone and increasing temperature zone. The variable temperate zone is the area of the earth that is thinner under the action of solar radiation. As the solar radiation energy cycle changes, changes in day and night temperatures can be observed in this region, as well as periodic changes over the course of a century or more each year. The annual temperature variation decreases with depth in accordance with a certain rule. Most of the world's plains, hilly areas of variable temperature zone thickness most of 15–20 m [4]. Solar radiation affects only the Earth's surface temperature, while heat diffusion in the Earth's crust affects both the deep surface temperature and the ocean temperature. Years of satellite observations have shown that changes in the sun's constant, about one in a thousand, are not enough to cause changes in the Earth's climate. How much warming is caused by the doubling of greenhouse gases (CO_2, CH_4, N_2O, etc.)? Tang [5] has used ancient historical data for rough estimation, the preliminary result is: doubled greenhouse gases cause warming $\leq 1°C$, this indicates the 1.5–4.5°C/2 × CO_2 simulation results obtained by various numerical models at home and abroad, at least without the support of ancient historical data. High temperature flow from the mantle is not only an important condition for the evolution of organic matter, but also a major factor in climate change [6]. The Earth's interior is constantly losing heat to the Earth's surface, in some places (such as near the crater or certain fault zones) in some cases (after an earthquake, for example) large values can be reached, it is called a "ground temperature sudden rise" or "geothermic anomaly" [7]. According to the monthly earth temperature data of Chinese stations from 1954 to 1985, a total of 70 stations were counted in the process of "ground temperature sudden rise", 53 "ground temperature sudden rise" were caused by "geothermal anomalies".

Research has shown that forced change of underlying surface is one of the most important causes of climate anomalies [8]. It is a well-known fact that tropical ocean temperature anomalies represented by El Niño events can cause global climate responses. The nature of climate change lies in its non-adiabatic nature [9]. Therefore, the abnormal heat condition of the underlying surface is an important cause of climate change. Alternating cold and warm atmospheric temperatures have certain similarities with ground temperature, and they have corresponding cold and warm centers. The change of ground temperature is synchronous with the change of atmospheric temperature, the difference is that the temperature changes in a small time scale more frequently than the earth temperature, and the earth temperature changes more energy in a larger time scale than the air temperature. The Earth's interior influences atmospheric processes by constantly sending matter and energy to the atmosphere, ultimately causing climate change [10].

To address climate change, a "pathogenesis" study of climate change must be performed because an accurate understanding of the pathogenesis of climate change is vital to understand current and future climate change.

2.1. Heat flow changes before and after fossil energy exploitation

Coal, oil, natural gas, shale gas and other fossil fuels are sensitive to temperature, pressure and other geological environmental factors. A series of physical, chemical, structural and structural changes of coal, oil and natural gas are inevitably caused by various tectonic events during geological evolution. Therefore, in the geological records of deep and shallow strata, it is necessary to include evidence related to the Earth's crust insulation seal failure and the surface heating caused by large-scale fossil fuel mining.

Li et al. [11] ground temperature changes observed in Huainan mining area, found that geothermal gradient of Huainan area (2.80–3.80°C/hm) is significantly higher than the old section (1.10–1.82°C/hm); the average geothermal gradient (3.42°C/hm) Pansan mining area west wing is higher than the east wing (3.14°C/hm), the east and west wings showed positive anomalies. In the old mining areas, the long-term mining activities have caused a lot of cracks in the rock strata, which makes cold water at the top infiltration, it cools the rock. In addition, the drainage system of the mine formed a cold water circulation system, which improved the cooling effect of the rock formation. After analyzing the abnormal causes, Li believes that the heat source causing the geothermal abnormality in Huainan mining area is mostly from within the earth, he made it clear that the thermal conductivity of coal is much lower than that of other sedimentary rocks. Therefore, in coal seam, especially in thick coal seam, high geothermal gradient will appear. The coal-bearing strata with more coal seams have more obvious thermal barrier effect on the whole than the depositional cover without coal. He and Wei [12] made an analysis of 30 boreholes and ground temperature test data of deep exploration and construction in Panzhihua coal mine, the general trend is that the high temperature area increases with depth, believed that its heat source comes from the residual heat of magma. Britain's deepest coal mine, in Lancashire, is about 1300 m deep. Verma measured original geothermal values of 26 of the 5 southern coal mines, geothermal gradient is 26.2°C/hm at a depth of 1220 m, the average original ground temperature is 45.7°C. Fourteen data points of coal seam in middle coal seam were measured, found due to the dense of coal seam, high thermal resistance, geothermal gradient of only 18.7°C/hm, thus get a lower thermal conductivity [13].

Xu et al. [14] the paleogeothermal study of Dagang Oilfield shows that the paleogeothermal effect in the oil and gas accumulation process of Dagang Oilfield produced obvious changes in the early stage that flattened at a later stage over time, this indicates that the paleogeothermal change is controlled by regional structure, and the peak area of paleogeothermal is consistent with the area with strong tectonic activity and frequent hydrothermal activity. Therefore, paleogeothermal change is closely related to regional tectonic environment and thermodynamic conditions. The period of paleogeothermal change is the period of intense crustal and structural movements, and also the peak of oil generation and oil and gas migration and accumulation. The heat flow statistics also show that the ground heat flow value of geological unit decreases with the increase of time in the last stage of tectonic thermal events [15]. According to Zhang et al. [16–19] research results, in the southern North China Basin groups of whole geothermal gradient between 13.0 and 39.9°C/km, average of 25.3°C/km. The ground heat flow value is between 30 and 89.6 mW/m^2, and the average is 53.7 mW/m^2. Compared with other geological units in mainland China, heat flow value is higher than that cryogenic basin of Tarim basin (44°C/hm) and Junggar basin (42.3°C/hm) in western China, and lower than that high temperature basin of the Bohai bay basin (69°C/hm) and Songliao basin (70°C/hm) in eastern China. From these data we can clearly see that unexploited oil and gas fields (Tarim Basin and Junggar Basin) have low heat flow value, large scale and longtime exploitation of oil and gas fields (Bohai Bay Basin, Dagang Oilfield, Songliao Basin, Daqing Oilfield, Jilin Oilfield and Liaohe Oilfield) with high heat flow value, oil and gas fields (Southern North China Basin–Southern North China Oil Field) between the two groups heat flow value is middle. On the basis of previous work, Qiu [20] according to a large number of rock thermal conductivity and thermal generation data, used heat conduction theory to calculate the deep temperature, analyzed the temperature distribution in the deep (below 4 km). The statistical average heat flow value from these heat flows get the Qaidam basin was 52.6 ± 9.6°C/hm. However, the heat flow value of the local wellhead is more than 70 mW/m^2, which is a thermal anomaly area of the basin. This is due to the large amount of oil production caused by the reduction of the heat insulation seal in the Earth's crust, increase in the earth heat flow.

Oil-bearing basins are rich in oil, gas, geothermal and other resources. Jiang et al. [21] used authigenic illite crystallinity and the chemical composition of authigenic chlorite to analyze the Jiyang depression Cenozoic ancient geothermal gradient. The results showed that the Jiyang depression Cenozoic ancient geothermal gradient is 37.2–38.2°C/hm. Gong et al. [22] using 703 drilling temperature measurement in Jiyang depression that nowadays the average geothermal gradient is 35.5°C/km, it is concluded that the paleogeothermal gradient is larger than the present geothermal gradient. Jiyang depression Zhanhua east block is located in Dongying estuary area, exploration proves that this area is a dual oil and gas accumulation area rich in oil and gas resources, with various oil and gas types. Based on *Ro* data of regional drilling temperature and vitrinite reflectance, Chen et al. [23] used the multi-stage thermal evolution model of lithosphere and basin scale, the present geothermal field in this area was analyzed and its thermal history was restored. The results are shown that (1) nowadays the geothermal gradient is 35.8°C/km, Gudao and Kendong areas geothermal gradient is higher, more than 37°C/km; (2) the early paleocene terrestrial heat flow value is 83.6°C/km, equivalent to the calorific value of a modern active rift. Since paleocene, the basin has shown a trend of gradual cooling. Although there have been two warming, the warming rate has finally decreased. The current terrestrial heat flow value is 63°C/km, the heat flow value is close to the global average; (3) the main source rocks in this area have undergone continuous heating and are now in the "oil generation window", in depth, there is a large hydrocarbon accumulation space, and the thermal evolution background is favorable for hydrocarbon generation. It can be seen that the formation of oil and gas effectively blocked the heat flow (**Table 1**).

2.2. Deep geotemperature and deep sea temperature both rise

Du et al. [24] analyzed the variation trend of the deep ground temperature of Lasa, indicating that the average ground temperature of Lasa has a significant upward trend in the past 45 years, the tendency to rate is 0.58–0.69°C/10a. Compared with the average increase rate of atmospheric temperature over the same period, geothermal temperature is growing even faster. In addition, geothermal observations from seasonal permafrost and weather stations in the permafrost regions of the former Soviet union show that the average annual geothermal temperature at most weather stations has increased over the past century [25]. In the Swiss Alps, the temperature of the permafrost layer below the surface has been increasing at a rate of 0.5–1.0°C/10a since 1980 [26]. The permafrost temperature measurement results, acquired in a north–south direction across Alaska, showed that the upper limit of the permafrost temperatures has increased by 0.5–1.5°C from the late 1980s to 1996 [27]. The Qinghai-Tibet Plateau permafrost temperature in the 1960–1990s increased by 0.2–0.3°C [28]. Qinghai-Tibet Railway in north and south of ground temperature linear heating rate is larger, especially in the south of the Qinghai-Tibet railway warming rate averaged 0.56°C/10a [29]. The permafrost ground

The Earth's crust component	Thermal conductivity (W/m K)
Coal	0.21
Petroleum	0.14
Natural gas	0.052
Oil shale	0.08
Shale gas	0.049
Combustible ice	0.121
Sedimentary rock	3.41
Granite	3.49
Basalt	2.17

Table 1. Thermal conductivity of various rocks and fossil fuels.

temperature in China's Daxinanling Amur region has increased 0.8°C from the 1970s to 1990s [30]. The ground temperature has increased 0.3–0.6°C in the Heilongjiang upper valley region from 1958 to 1990 [31]. Observations based on the Qinghai-Tibet highway and Railway geothermal features and degradation mode by Jin et al. and Li et al. [32, 33], showed permafrost degradation, ground temperature increases, the summer's biggest melt depth deepening, winter freezing depth decreases, and permafrost thickness thinning, or disappear completely in some areas. At present, the downward melting rate of frozen soil in Qinghai-Tibet Railway is about 6–25 cm a^{-1}, while the upward melting rate reaches 12–30 cm a^{-1}. The annual average temperature and ground temperature warming rate are 0.33 and 0.37°C/10a, respectively. In general, ground temperature rises faster than air temperature.

According to a report in the British NEW SCIENTIST on 12 Dec 1994, climate warming is not consistent with climate change expectations based on the accumulation of greenhouse gases in the atmosphere on earth. Researchers believe the southwest Pacific is a valuable reference for monitoring climate change because it has fewer cities and less air pollution. New Zealand's National Institute of Water and Atmospheric Research has provided first-hand information on the warming of the Indian Ocean. The Antarctic Climate and Ecosystem Cooperative Research Centre (Australia Tasmania Hobart) of Nathan Bindoff to early and mid-1960s records of ships across the Indian Ocean temperature data and research ship 1987 Darwin recorded data are compared, and he calculated in latitude 32° south 250–1500 m in the depths of the ocean temperature rose about 0.5°C. Bindoff believes that temperature changes in the deep ocean are an important indicator of global climate change. He suggests that when measuring temperatures deep in the ocean, seasonal fluctuations are small. Thus, measurements of the deep ocean can provide more accurate results and fewer observations than measurements of sea level. The Indian Ocean has proved to be the third ocean in which deep water is warming. Bindoff published similar results in 1992 and showed that temperatures in the southwest Pacific increased at almost the same rate. Gregorlo Parrilla and his team at the Spanish institute of oceanography found that the north Atlantic was also warming [34].

Professor Pollack [35] after analyzing more than 60 geothermal data in South Africa found that, in the past 100 years in the area ground temperature increased by 0.3–0.8°C, an average of 0.55°C, completely consistent with the results of global change research. International famous geothermal scientists and members of the National Academy of Sciences Arthur H. Lachenbruch research of a lot of temperature data from northern Alaska drilling (inside the Arctic Circle) and came to the conclusion that this area has increased 2–4°C temperature over the past century [36]. After studying more than 30 borehole temperature measurements from Cuba, vice chairman of the international heat flow committee and that year director of the Institute of Physics of the Czechoslovak Academy of Sciences, Dr. Čermák pointed out that, Cuba region increased temperatures 2–3°C in the past 200–300 years [37]. Professor Mareschal from the Université du Québec, Canada and Dr. Jessop from the Geological Survey Institute of Canada [38] based on a large amount of temperature measurement data in central and eastern Canada, reported temperature increase 1–2°C in the past 100–200 years, and most of the temperature changes inferred from ground temperature data are consistent with the observation results of meteorological stations.

3. Energy mining causing all kinds of disasters

In the space of just 3 years from 1998 to 2000, four curves representing temperature changes in the northern hemisphere or the world over the last 1000 years have been published internationally. Why extend the study to 1000 years, and why build a temperature curve in the northern hemisphere or around the world? There are two main reasons: first, long enough sequences to show whether warming in the twentieth century was abnormal and thus whether it was the result of human activity. Second, determine whether Medieval Warm Period (MWP) and Little Ice Age (LIA) really exist in the last millennium from the northern hemisphere or global scale. Because both events occurred before human activity could have a significant impact, most authors attribute them to natural climate change. If natural change is also global, and the magnitude of change is close to, or even greater than, the warming of the twentieth century, it suggests that the warming of the twentieth century may also be caused by natural causes. Wang et al. [39] synthetically analyzed four temperature sequences established by Mann et al., Jones et al., Crowley et al., and Briffa that represent the average temperature of the northern hemisphere or the global in the last 1000 years, and using the 30 sites information for nearly 1000 years of global average temperature sequence (W), and USES the energy balance model for nearly 1000 years of the simulation results of temperature change (S), compared with the simulation results on the various temperature sequence, the conclusion is that Little Ice Age is relatively obvious, and Medieval Warm Period is not as consistent as that of Little Ice Age. Calculate the centennial average in 1925, 1950, 1975 and 2000, the 1000 average anomaly at 0.50°C or so, the centenary average is significantly higher than the average for any century from the eleventh to the twelfth century. It is clear that climate change over the past century or so has not been caused by natural causes. Analysis of the insulation sealing function of fossil energy, overexploitation, depth of surface temperature increase of the surface heat flux and ocean temperature rise, it can explain the relationship between environmental change and various abnormal disasters in recent 100 years. For example, long-term continuous rise of the Earth's surface temperature in the deep and shallow layers changes the mechanical structure of the Earth's crust, soft change of soil and rock cohesion, landslides, debris flow and other geological disasters will occur frequently. The increase of ground temperature causes harmful substances in rocks to dissolve into groundwater, some areas will face water quality induced water shortage.

Scientists observed that in the last 50 years nearly 40 offshore areas around the world have become dead seas, and the main reasons are pollution and climate warming. A study by scientists from China's state oceanic administration confirmed that sea levels along the Pacific west coast are expected to accelerate rise in the future, it will rise by 10–40 cm in 2030 and 40–90 cm in 2100.

Underlying surface and sea water temperature rise, not only led type on glaciers and permafrost to melt or even disappear, will also make the water acid is higher and higher, the survival of Marine life is more and more difficult, Marine biological extinction, and global warming is getting worse. Lake water temperature rises, lake water eutrophication aggravates, blue algae, red tide frequently appears, water quality deteriorates. Ground temperature rise, snow mountains melt, and snow line rises, droughts and floods occur frequently and land and ocean degradation aggravate.

Time	Coal	Oil	Gas	ENSO frequency of occurrence
1649–1879				Closely related to submarine volcanic eruption.
1880–1980	1500 (1500)	517 (517)	30 (30)	Happen once between 2 and 7 years, duration is about 1 year.
1981–2005	1125 (2625)	440 (957)	20 (50)	The interval of occurrence is about 3 years and the duration is about 15 months, several El Nino events that have occurred since the 1990s have taken only about half a year apart, the longest interval is less than 2 years, lasted 3 years.

Table 2. Exploitation quantity of global coal, crude oil and natural gas vs ENSO frequency of occurrence output (add up) (unit: billion ton, trillion m^2).

Higher sea temperatures in the deep and shallow layers allow water in the ocean to evaporate more rapidly, increase rainfall in flood-prone areas. The winter became warmer and the stock of snow decreased. Melting snow water no longer trickles down mountains for months, but flows directly into rivers. More and more rivers are turning into seasonal rivers, while a once-in-a-century flood are now occurring every year. Too much heat enters the atmosphere from the earth's interior, causes the subtropical high and cold air intensity to increase. It's bound to make summer hotter and much colder in winter. The frequent of ENSO and La-Niña not only cause the abnormal climate, but also causes the extreme rain and drought due to the disorder of precipitation. The world's climate will gradually become polarized from summer to winter, from rainy to dry season.

According to statistics, the number of natural disasters worldwide has more than tripled in the past 20 years. The global average of 120 natural disasters per year in the early 1980s has risen to about 500 now. Climate change and various environmental and geological disasters have been reported (**Table 2**) [40–45].

The increase of underlying surface temperature causes the distribution of tropical plants to move northward. The invasion of harmful species will have a significant impact on the distribution of plants in the north. Rising temperatures have led to a decline in sperm counts in male animals, the degeneration of male genetic material in plants and the mass extinction of animal and plant populations. The ground temperature, water temperature, ocean temperature and atmospheric temperature rise, have made local tropical pests and diseases, such as the south of schistosomiasis, malaria and cockroaches carry diseases) mass migration north, the distribution of threat to human health. Frequent warm winters have caused changes in the biological habits of nature, such as hibernation. Due to the heavy use of highly toxic pesticides, the propagation of pests and the multiplication of environmental toxicity will exacerbate environmental degradation.

4. Even if human stop GHGs, change continue for 1000 years

Is the earth a giant fireball with high temperature and pressure inside? Do fossil fuels such as coal, oil and gas provide efficient, long-lasting thermal insulation to the crust? Are hard crustal rock layers insulated? The Mawangdui Han Dynasty Tomb in Changsha and the Ming Dynasty Ding Tomb in Beijing both have inadvertently conducted objective, long-term "scientific experiments" on the thermal insulation of charcoal, grease and rock, there is a sharp contrast. Mawangdui Han Dynasty Tomb did not use stone materials, but the tomb was sealed with charcoal and white paste mud containing grease, after more than 2100 years, all the articles in the tomb, including coffins and bodies, including silks and grain, are well preserved [46, 47]. By contrast, in the Ming Dynasty Ding Tomb, the six hardest layers of white marble and striped stone were used to build tombs and coffins, but the owners were not preserved. In less than 400 years the body of the Wanli Emperor had all rotted away, leaving only a dry skeleton [48, 49]. Therefore, no matter how thick it is, no matter how hard the rock is, it is impossible to resist the normal earth heat flow, let alone the increased heat flow. This fully shows that coal and oil have high efficiency and lasting thermal insulation. If the combination of white plaster and charcoal can seal the Mawangdui Han Dynasty Tomb perfectly, then the combination of oil, gas and coal can seal the Earth's crust perfectly.

Since the industrial revolution in Britain, the massive exploitation of countless coal mines, oil fields and gas fields has caused tens of millions of parts to be superimposed into a whole effect. Global changes caused by "heating" of the Earth's crust and underlying surfaces are already evident.

The relationship between atmospheric CO_2 concentration and climate has been challenged by the fact that recent temperature increases have been much lower than scientists predicted for CO_2 concentration increases. The theory of the greenhouse effect of climate change is perplexed by a large number of climate and natural anomalies [50–53]. In fact, most of the CO_2 entering the air is absorbed by the sea water and gradually becomes carbonate deposited on the seabed, forming rocks, or move to land through the shells, bones and dust of aquatic creatures. Carbonate absorbs CO_2 from the air and becomes bicarbonate, which is dissolved in water and eventually returned to the ocean [54]. With the continuous absorption of heat from the underlying surface and sea water, the temperature increase rate of the ground and sea temperature will be significantly greater than the average temperature increase of the atmosphere [55]. At some point, it might even happen the global average temperature will stop rising.

Research by Amanda Scott of NOAA and others shows that global warming is irreversible and accelerating. Even if humans stop emitting greenhouse gases, warming will continue for 1000 years. This conclusion refutes the theory of greenhouse effect of climate change, suggesting that there are other reasons for global warming.

Upward melting of glaciers and frozen soils [32, 56] (upward, bottom up melting) and deep sea temperatures and ground temperatures are increasing are far greater than the magnitude of increase of the average atmospheric temperature for the same period [34, 53, 57–58], and these are things that the theory of greenhouse effect cannot explain.

To sum up, there is the following evidence to support that the Earth's crust heat insulation seal damage caused by fossil energy exploitation is the main cause of global change: (a) Global warming is not consistent with climate change as predicted by the buildup of greenhouse gases in the Earth's atmosphere. (b) Global warming is irreversible and accelerating, even if humans stop emitting greenhouse gases, warming will continue for 1000 years. (c) Scientists have shown that high heat flow from the mantle is not only an important factor in the evolution of organic matter, but also a major factor in climate change. (d) The Earth's interior influences atmospheric processes by constantly sending matter and energy to the atmosphere, ultimately causing climate change. (e) The formation of fossil energy in the Earth's crust effectively blocked the Earth's heat flow. (f) According to scientific observation, after the exploitation of fossil energy, the earth heat flow can reach a very large value, and the phenomenon of "ground temperature sudden rise" occurs. The earth and sea temperatures in the deep and shallow layers have increased significantly. (g) The excavation of the Mawangdui Han Dynasty Tomb indicates that the sealed with charcoal and white paste mud containing grease can effectively block the earth heat flow for more than 2100 years, and the Ming Dynasty Ding Tomb, built from six layers of extremely hard rock, has no thermal insulation. (h) The fossil energy in the crust has obvious thermal resistance [59].

We have every reason to believe that, once the human society attention to this problem, there must be many scientists in the process of their practice to find out more and more due to the endless exploitation of the human mass on the fossil energy of the Earth's crust insulated sealing damage, increase the ground temperature, SST caused cause ecological geological disaster, climate change, and a series of direct or indirect evidence of meteorological disasters.

Acknowledgements

Data supporting this article come from China's Ministry of Energy. Data cannot be released because of national security concerns.

Author details

Yao Mu and Xinzhi Mu*

*Address all correspondence to: 92690_fan@sina.com

BiboShenzhen 518 Institute, Shanghai Zhangjiang Hi. Tech. Park, Shanghai, China

References

[1] Tang M. The impact of lithosphere forcing on climate change. Scientia Meteorologica Sinica. 1995;**15**(4):2-6 (in Chinese with Abstract in English)

[2] Tang M, Gao X. A rule and three guesses about the global mean (zero-dimension) climatic variation. Journal of Nanjing Institute of Meteorology. 1992;**15**(1):17-21 (in Chinese with Abstract in English)

[3] Tang M, Jian Z. Seasonal mean soil temperature anomaly field at depth 3.2 m and its application in prediction for flood season. Plateau Meteorology. 1994;**13**(2):178-187 (in Chinese with Abstract in English)

[4] Wang J, Huang S, Huang G. Chinese Geotemperature Distribution Basic Characteristic. Beijing: Seism Press; 1990. pp. 59-60 (in Chinese)

[5] Tang M, Zhu D, Gao X. Review and prospect of the research of earth system's evolution. Advances in Earth Sciences. 2004;**19**(1):55-62 (in Chinese with Abstract in English)

[6] Shi D, Liu W, Zheng J. Theory analysis on deep seated gas and its potential study. Advance in Earth Sciences. 2003;**18**(2):236-244 (in Chinese with Abstract in English)

[7] Tang M, Zhang J. Approach on the underground suddenly warming relative to short-term climatic variations. Plateau Meteorology. 1990;**9**(4):364-370 (in Chinese with Abstract in English)

[8] Shukla J. Predictability of time averages, Part II: The influence of the boundary forcing. In: Burridge DM, Kallen E, editors. Problems and Prospects in Long and Medium Range Weather Forecasting. New York: Springer-Verlag; 1984. pp. 155-206

[9] Huang R. Physics Foundation of Short-Term Climate Change. Beijing: Meteorology Press; 1991. pp. 88-112 (in Chinese)

[10] Gao X, Tang M, Zhu D. Some thoughts on climate system and earth system. Chinese Journal of Geophysics. 2004;**47**(2):364-368 (in Chinese with Abstract in English)

[11] Li H, Zhu Y, Yi J. Analysis of geotemperature change law and abnormal factors in Southern Anhui. Coal Mine Safety. 2007;**11**:68-71 (in Chinese)

[12] He G, Wei K. Geothermal study in Baoding mining area, Panzhihua. Coal Geology of China. 2009;**19**(5):51-53, 81 (in Chinese with Abstract in English)

[13] Shaolin LV. Primitive geotemperature survey studying in western British mining area. Coal Mine Safety & Research. 1985;**2**:51-54 (in Chinese)

[14] Xu X, Zou H, Gao F, Yang Y, Sun X, Xiao D, Wu Z. The characteristics and evolution of the paleo-temperature during formation process of oil-gas in Dagang Exploration Area. Journal of Jilin University (Earth Science Edition). 2003;**33**(4):457-463 (in Chinese with Abstract in English)

[15] Pollack HN, Hurter SJ, Johnson JR. Heat flow from the Earth's interior: Analysis of the global data set. Geophysical Reviews. 1993;**31**:267-280

[16] Zhang P, Wang L, Liu S, Li C, Ding Z. Geothermal field in the south Huabei basins. Progress in Geophysics. 2007;**22**(2):604-608 (in Chinese with Abstract in English)

[17] Wang J, Wang J'a, Shen J, Qiu N. Heat flow in Tarim basins. Earth Science-Journal of China University of Geosciences. 1995;**20**(4):399-404 (in Chinese with Abstract in English)

[18] Wang S, Hu S, Li T, Wang J, Zhao W. Heat flow in Junggar basins. Science Bulletin. 2000;**45**(12):1327-1332 (in Chinese)

[19] Hu SB, He LJ, Wang JY. Heat flow in the continental area of China: A new data set. Earth and Planetary Science Letters. 2000;**179**:407-419

[20] Qiu N. Research on heat flow and temperature distribution of the Qaidam Basin. Journal of China University of Mining & Technology. 2001;**30**(4):412-415 (in Chinese with Abstract in English)

[21] Jiang H, Xiao Y, Zhou L. Analysis of Cenozoic subsurface temperatures of the Jiyang depression. Geology in China. 2008;**35**(2):273-278 (in Chinese with Abstract in English)

[22] Gong Y, Wang L, Liu S, Guo L, Cai J. Distribution characteristics of temperature field in Jiyang depression, Shandong, North China. Chinese Journal of Geophysics. 2003;**46**(5):652-658 (in Chinese with Abstract in English)

[23] Cheng B, Xu L, Xiang X, Mu X. Present-day geothermal field and thermal history of the Zhanhuadong block, Jiyang, depression. Chinese Journal of Geophysics. 2001;**44**(2):238-244 (in Chinese with Abstract in English)

[24] Du J, Hu J, Yang Y, Lhak P, Lu H. Variations trend of soil temperature at deep layers in Lhasa from 1961 to 2005. Journal of Applied Meteorological Science. 2008;**19**(1):96-100 (in Chinese with Abstract in English)

[25] Gilichinsky DA, Barry RG, Bykhovets SS. A century of temperature observations of soil climate: Methods of analysis and long-term trends. In: Lewkowicz AG, Allard M, editors. Proceedings of the Seventh International Conference on Permafrost. Yellowknife, Canada. 1998. pp. 313-317

[26] Haeberli W, Hoclzle M, Keller F. Monitoring the long-term revolution of mountain permafrost in the Swiss alps. In: Proceedings of Sixth International Conference on Permafrost. Vol. 1. Guangzhou: South China University of Technology Press; 1993. pp. 214-219

[27] Osterkamp TE, Romanovsky VE. Evidence for warming and thawing of discontinuous permafrost in Alaska. Permafrost and Periglacial Processes. 1999;**10**(1):17-37

[28] Wang S. Permafrost changes along the Qinghai-Xizang highway during the last decades. Arid Land Geography. 1993;**16**(1):1-7 (in Chinese with Abstract in English)

[29] Li D, Liu M, Zhong H, Wu Q. Interdecadal change trend of surface air and ground temperatures along Qinghai-Xizang railway and relationship between the change and terrain. Plateau Meteorology. 2005;**24**(5):694-699 (in Chinese with Abstract in English)

[30] Gu Z, Zhou Y, Liang F. Permafrost features and their changes in Amur area, Daxinganling prefecture, northeastern China. In: Proceedings of Sixth International Conference on Permafrost. Vol. 1. Guangzhou: South China University of Technology Press; 1993. pp. 204-209 (in Chinese)

[31] Yu S, Wang Z. The degeneration of permafrost in Upstream Valley region of Heilong River. In: Proceedings of Sixth International Conference on Permafrost. Vol. 1. Guangzhou: South China University of Technology Press; 1993. pp. 755-757 (in Chinese)

[32] Jin H, Zhao L, Wang S, Jin R. Geotemperature characteristics of permafrost and degenerate way along Qinghai-Xizang highway. Science in China Series D Earth Sciences. 2006;**36**(11):1009-1019 (in Chinese)

[33] Li D, Zhong H, Wu Q, Zhang Y, Hou Y, Tang M. Analyses on changes of surface temperature over Qinghai-Xizang plateau. Plateau Meteorology. 2005;**24**(3):291-298 (in Chinese with Abstract in English)

[34] Xiao Q. Global warming is explained temperature changes in the depth of the ocean. Science and Technology Review. 1995;**5**:54 (in Chinese)

[35] Pollack HN. Surface temperature history and subsurface temperature profiles: A comparison from Southern Africa. In: IUGG XX General Assembly, IASPEI Program & Abstract. 1991. p. 238

[36] Lachenbruch AH, Marshall BV. Changing climate: Geoheat evidence from permafrost in Alaskan Arctic. Science. 1991;**234**:689-696

[37] Čermák V, Bodri L, Šafanda J. Underground temperature and climate of the past 300 Years: Evidence from Cuba. In: IUGG XX General Assembly, IASPEI Program & Abstract. 1991. p. 241

[38] Jessop AM, Hugo B, Mareschal JC. Recent warming in eastern and Central Canada inferred from borehole temperature measurements. In: Proceedings American Geophysical Union Fall Meeting; San Francisco. 1991. p. 241

[39] Wang S, Xie Z, Cai J, Zhu J, Gong D. Global average temperature change research nearly one thousand years. Progress in Natural Science. 2002;**12**(11):1145-1147 (in Chinese)

[40] Steffensen JP, Andersen KK, Bigler M, Ausen HB, Dahl-Jensen D, Fischer H, et al. High-resolution Greenland ice core data show abrupt climate change happens in few years. Science. 2008;**321**:680-684

[41] Fujita K. Another Antarctic rhythm. Nature. 2011;**471**:45-46

[42] Petit RJ, Hu FS, Dick CW. Forests of the past: A window to future changes. Science. 2008;**320**:1450-1452

[43] Visser ME. Fatter marmots on the rise. Nature. 2010;**466**:445-446

[44] Pörtner HO, Farrell AP. Physiology and climate change. Science. 2008;**322**:690-692

[45] Marin A. Riders under storms: Contributions of nomadic herders' observations to analysing climate change in Mongolia. Global Environmental Change. 2010;**20**:162-176

[46] Xiong C. A record of Mawangdui Han dynasty tomb excavate. Nature Explore. 2005;**4**:64-71 (in Chinese)

[47] Xiong C. She walked us from 2000 years ago-on probing the secrets of the Mawangdui Han dynasty tomb female corpse. Nature Explore. 2005;**4**:72-76 (in Chinese)

[48] Ni F. Puzzles behind the exhumation of Beijing Ming dynasty Ding mausoleum. Cultural Dialogue. 2008;**9**:55-58 (in Chinese)

[49] Wang Q. Revealing the secrets of Beijing Ming Dynasty Ding mausoleum. Age of Culture and History. 2003;**12**:47-53 (in Chinese)

[50] Doug S. Has global warming stalled? Nature Climate Change. 2013;**3**:618-619

[51] Guemas V, Doblas-Reyes FJ, Andreu-Burillo I, Asif M. Retrospective prediction of the global warming slowdown in the past decade. Nature Climate Change. 2013;**3**:649-653

[52] Crucifix M. Climate's astronomical sensors. Nature. 2008;**456**:47-48

[53] Zahn R. Beyond the CO_2 connection. Nature. 2009;**460**:335-336

[54] Riebesell U. Acid test for marine biodiversity. Nature. 2008;**454**:46-47

[55] Rosenthal Y, Linsley BK, Oppo DW. Pacific Ocean heat content during the past 10,000 years. Science. 2013;**342**:617-621

[56] Kaus BJP. Heating glaciers from below. Nature Geoscience. 2013;**6**:683-684

[57] Sloyan BM, Wijffels SE, Tilbrook AR, Atsuro K, Atsumata K, Urata AM, Macdonald AM. Deep Ocean changes near the western boundary of the South Pacific Ocean. Journal of Physical Oceanography. 2013;**43**(10):2132-2141

[58] Cartes JE, Fanelli E, Lloris D, Matallanas J. Effect of environmental variations on sharks and other top predators in the deep Mediterranean Sea over the last 60 years. Climate Research. 2013;**55**:239-251

[59] Wang J, Sun Z. Terrestrial Heat. Beijing: Tsinghua University Press; 2001. pp. 5-97 (in Chinese)

Chapter 9

Impact of Cold Wave on Vulnerable People of Tarai Region, Nepal

Bandana Pradhan, Puspa Sharma and
Pushkar K. Pradhan

Additional information is available at the end of the chapter

http://dx.doi.org/10.5772/intechopen.82201

Abstract

Climate extremity phenomena are increasing with the global climate change. Cold wave is one of these climate extremities affecting the health of people, especially vulnerable groups. Nepal is also experiencing the impacts of global warming on its temperature patterns. The climate data of more than four decades have shown an increasing trend of annual temperatures across Nepal. However, the change in temperatures is found varying greatly among its three broad physiographic regions: Tarai, hill, and mountains, as well as among four distinct seasons: winter, pre-monsoon, monsoon, and post-monsoon during a year. Further, since the last two decades Nepal has experienced climatic extremities such as heat wave, cold wave, precipitation concentration, prolonged dryness affecting livelihood of the people and demographic features like mortality, morbidity, etc. This study intends to deal with the impact of cold extremity on the death of vulnerable people such as children and elderly in the Tarai region. It draws on meteorological data for four decades since 1974. The magnitude of mortality rate of those vulnerable people is analyzed from 1974 to 2013, and prediction of mortality rate is made with respect to decrease in temperature or intensity of cold wave.

Keywords: cold wave, temperature change, vulnerable people, number of death, Tarai region, Nepal

1. Introduction

The evidences of impacts of climate change across the world are that there has been an increase in climate extremity phenomena such as cold wave, heat wave, extreme depressions, intense precipitation, rising number of warm days and nights, and decreasing number of cold days and nights. AR4 and AR5 IPCC have predicted an increase in frequency and intensity of warm spells or heat waves in the end of twenty-first century, affecting to increase vector-borne diseases, water-borne diseases, reduce cold-related mortality, and diminish food production and labor productivity at different levels over most land areas of the earth [1, 2]. As a matter of fact, there is large number of studies on health effects of heat waves [3, 4]. Some of the studies argue that the contribution of rising in minimum temperatures has decreased in number of deaths associated with cold spells [5, 6]. On the other hand, there are few studies dealing with cold spells and health impacts. For instance, some studies indicated that the effects of extreme cold temperatures are generally more prolonged than heat wave without mortality displacement [3, 7]. However, most of the existing studies on health effects of cold spells are found to be associated with the temperate climate regions in developed countries, while there are very few such studies carried out in tropical or subtropical regions of developing countries [3, 8–11].

Nepal has experienced global warming and its impacts on forming climate extremities, ill-health of the people, change in agricultural production patterns, etc. over the past recent decades. Cold wave is one of the climate extremities due to global warming in Nepal. The studies of National Agriculture Research Council (NARC) have indicated negative impacts of cold wave on agricultural productivity in Nepal [12]. Other studies have shown the health of the inhabitants being affected due to cold wave in the Tarai region of Nepal in the last two decades [13–16]. The present chapter intends to analyze the climate change patterns and the climate extremities such as cold wave and its impacts on the vulnerable populations in the Tarai region of Nepal.

2. Methodology

The vulnerable population is defined in terms of age group such as children below 5 years of age, pregnant women, and elderly population above 65 years of age [17]. The three subsets of under-five children, such as neonates <1 months, infant <1 year and <5 years, of which neonates is the most vulnerable and it is followed by other subsets [18].

The climate prevailing in Nepal can be divided into four seasons, based on rainfall and temperature conditions. They are rainy summer or Monsoon (June–September with rainy, hot, and humid weather), winter (December–February with coldest and driest weather), pre-monsoon (March–May with hot weather and thunderstorms) and post-monsoon (October–November with cool and pleasant weather). The climate data including monthly minimum and maximum temperatures for all individual years from 1974 to 2014 by the physiographic regions, such as mountain, hill, and Tarai have been acquired from all 67 weather stations from the Department of Hydrology and Meteorology, Kathmandu, and Nepal [19]. These data have been used for describing climate change patterns for all physiographic regions across the country in general and for analyzing seasonal trends of climate and climatic extremities for the Tarai region in particular. The prediction of trends of temperatures by year has been carried out for two distinct slots: 1974–2014 and 2000–2014.

Data on daily death of the vulnerable population groups due to cold wave during the winter season from 1974 to 2013 for all districts of the Tarai region were obtained from available sources [18–22]. The contribution of seasonal temperature change, mainly, cold wave to the deaths of the vulnerable groups, and the mortality rate have been analyzed by using multiple regression analysis.

The multiple linear regression analysis has been used to develop a model for predicting mortality from the climatic variables at different time lags. This relationship is given by the equation [23]:

$$Y = \beta o + \beta 1X1 + \beta 2X2 \ldots\ldots\ldots \beta kXk + \varepsilon \qquad (1)$$

where β_k is coefficients, X_i is the predictor, Y is mortality (predicted), β_o is a constant and ε is random error.

The perception about the death due to cold spell or wave and status of vulnerable population groups were carried out by informal talking among 25 respondents selected randomly: 5 each from different walks of life such as the local communities, government personnel, public health personnel, female community health volunteers and school teachers.

3. Findings

3.1. Brief introduction to physiography, climate, and population of Nepal

3.1.1. *Physiography*

Geographically, Nepal can be divided into three broad physiographic regions, namely mountain, hill, and Tarai from north to south (**Figure 1**). The altitudes of these three regions range from 8848 m above sea level (masl) in the north to 60 masl in the south over an average north-south span of 193 km [24]. Tarai refers to plain topography in Nepal.

The Tarai is the smallest physiographic region, sharing 23% of the country's total area, but it has the largest population with over 50% of the nation's total population of 26.6 million (**Table 1**). Population has increased consistently in this region during the past decades. In 1971, the Tarai's population had shared nearly 38% of the country's total population that increased to over 50% in 2011 [17]. The rapid growth of the Tarai population is considered due to natural cause and other causes including internal migration of population from the hills and international migration from adjoining Indian states [17, 21].

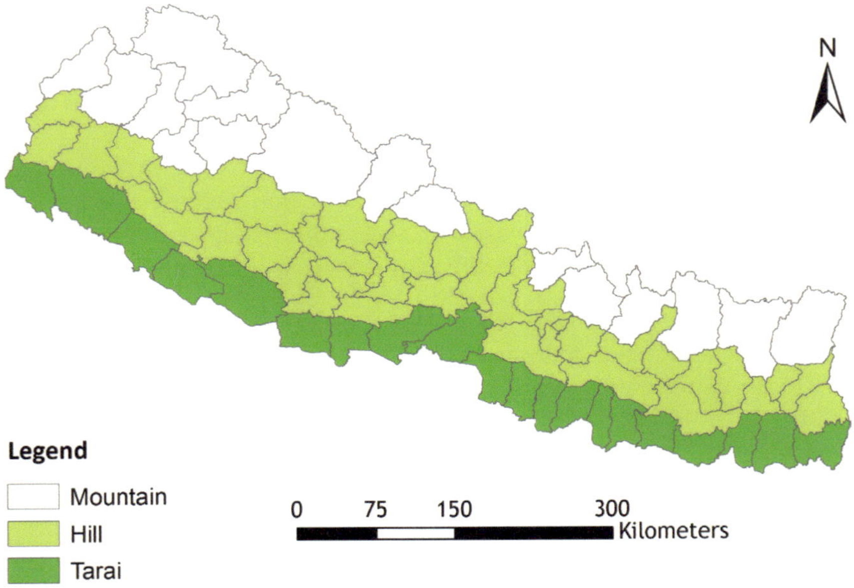

Figure 1. Physiographic regions of Nepal.

Ecological regions	Approximate elevation (m)	Area		Number of districts	Population (2011)	
		km²	%		In million	%
Mountain	>4870	51,817	35.2	16	1.78	6.7
Hill	300–4870	61,345	41.7	39	11.39	43.0
Tarai	<300	34,019	23.1	20	13.32	50.3
Total		147,181	100	75	26.49	100

Source: [17].

Table 1. Broad physiographic regions and their features.

Table 2 exhibits that the combined populations of the vulnerable groups (under-five children, pregnant women, and elderly) account for about 17% of the Tarai's total population. For Nepal, the life expectancy at birth of Nepalese population is 66.51 years, whereas the death rate is 6.75 deaths/1000 population and infant mortality rate is 32 deaths/1000 live births [18].

3.1.2. Climate

Nepal lies within the subtropical climatic zone over the globe [19]. The climate is largely influenced by the Monsoon system, but there is also an influence of the cyclonic system originating from the Mediterranean Sea during the winter season. Owing to rise of altitude of mountains considerably from the south to the north, Nepal possesses diverse climate types ranging from

Vulnerable groups	Total population	% of vulnerable population
Under-five population	1,380,169	10.3
Expected pregnancies	350,497	2.6
Elderly population	534,018	4.0
Total	2,264,684	16.9

Source: [17, 21].

Table 2. Distribution of population of vulnerable groups in Tarai region, Nepal.

sub-tropical in the Tarai region to the arctic in the high Himalayas. Likewise, the annual precipitation also ranges from over 5000 mm in the Western and Eastern midland mountains to below 150 mm in the northern areas beyond high Himalayas, with an annual mean precipitation of 1858 mm [19, 25].

In Nepal, the annual trends of temperature patterns vary remarkably among four different seasons: summer monsoon, post-monsoon, winter and pre-monsoon, and those three physiographic regions (**Table 3**). The annual maximum and minimum temperature trends for the country as a whole are 0.056 and 0.002°C/year, respectively [19]. **Table 3** shows a negative trend of maximum temperature in contrast to the positive trend of minimum temperature during the winter season in the Tarai region [19].

The mean annual maximum temperature for the Tarai region is 30.4 at 95% confidence interval (CI) of 36–30 and mean annual minimum temperature is 18.3 at 95% CI of 16–25. During the winter season, the mean minimum temperature in the Tarai region remains at 9.8°C with 95% CI of 9.5–10.1°C and mean maximum temperature is 23.2 with 95% CI of 22.7–23.7°C. While analysis of temperature trends by year is performed, a conspicuous distinction is found between two slots of years, such as 1974–2014 and 2000–2014 (**Table 4**). During the second slot of years: 2000–2014, negative trends are found in the annual mean minimum temperature, as well as in both mean maximum and minimum temperatures in the winter season in the Tarai region, but found a positive trend in the annual mean maximum temperature in this slot of years. Conversely, positive trends are found in all three temperature conditions during the first slot of years: 1974–2014 except in the winter maximum temperature, which shows negative trend. Thus, the analysis of two slots of periods of years shows a decreasing temperature scenario, particularly during the winter season in the Tarai region.

3.2. Cold wave impacts on Tarai people

Cold waves generally occur in the Tarai region from mid-November to mid-February. On average, the duration of cold waves is found to be 8 days. In 2003, the duration of cold waves remained to be up to 60 days. However, the duration of cold waves prevailing in the Tarai has risen since 2004, compared to that in 2000 (**Figure 2**).

Record of hourly average temperature data shows that the peak cold temperature appears to remain from December to January, where minimum temperature goes below 5°C for few hours during night (**Figure 3**).

Physiographic regions	Temperature trends °C per year (1974–2014)				
	Winter	Pre-monsoon	Monsoon	Post-monsoon	Annual
Maximum temperature					
Tarai	−0.004	0.018	0.036	0.028	0.021
Hill	0.046	0.049	0.055	0.052	0.052
Mountain	0.101	0.076	0.072	0.085	0.086
Minimum temperature					
Tarai	0.025	0.015	0.015	0.013	0.018
Hill	0.004	0.004	0.014	0.006	0.010
Mountain	−0.056	−0.021	0.013	−0.025	−0.015

Table 3. Seasonal temperature trends by physiographic regions, Nepal.

Temperature conditions (°C/year)	1974–2014	2000–2014
Annual maximum	0.021	0.031
Annual minimum	0.018	−0.040
Winter maximum	−0.016	−0.062
Winter minimum	0.024	−0,024
Source: [19].		

Table 4. Trends of maximum and minimum temperature trends in the Tarai region.

There are altogether 30 different types of disaster events being recorded in Nepal [20]. Of these events so far recorded, cold wave is the most crucial one. It is as large and serious as damage of crops due to disaster (**Table 5**). The effect of the cold wave is found across the country (see also **Figure 7**), or primarily in the high mountain region, where there is cold in most of the year, which is, however, not so significant because there exists very thinly scattered population, which mostly have been adapted to the cold climate. But it is crucially very significant in the Tarai region of Nepal, as it possesses largest population size and its poverty level is comparably large [17].

Figure 4 exhibits yearly trend of deaths of people due to cold wave, which is found to be increased exponentially at 13% per year. It is found that the number of deaths due to cold wave has speedily increased, particularly from the year 2000 onwards.

The magnitude of deaths of people due to cold wave in the Tarai is comparably large in the country, as shown in **Figure 5**. The total deaths from cold wave from 1974 to 2013 were recorded at 822. Of these total deaths, 89% took place in the Tarai region, followed by 9 and 2% in the Hill and the Mountain regions, respectively [20, 22].

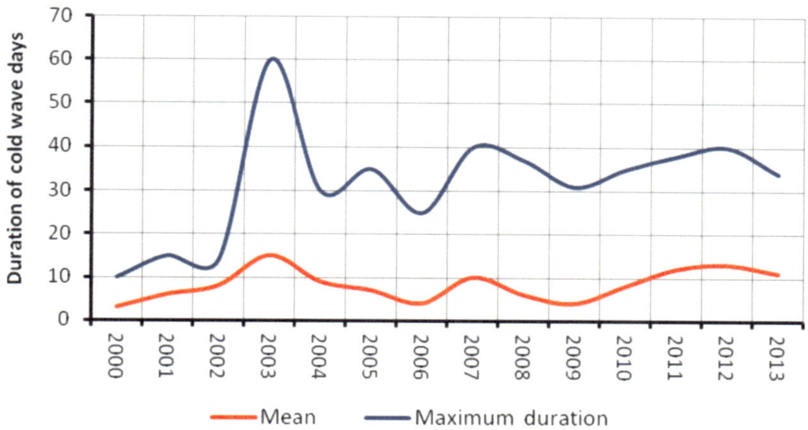

Figure 2. Trend of duration of cold waves in Tarai region, 2000–2013.

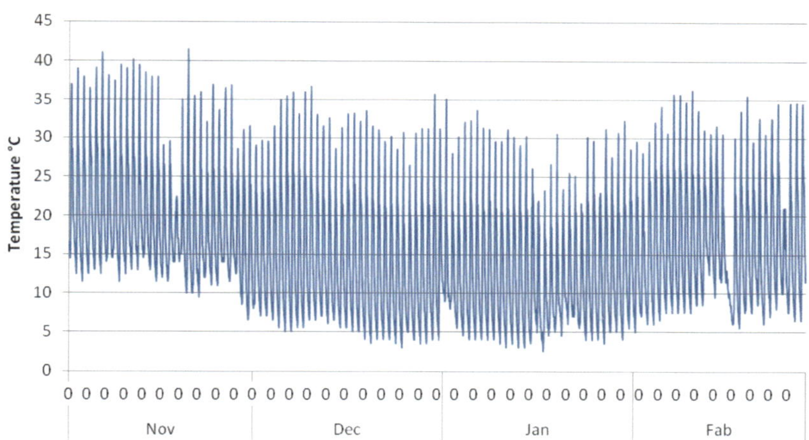

Figure 3. Distribution of average hourly temperature by winter months, Tarai region.

The distribution of average number of deaths due to cold wave by district across the country is found to be varied remarkably. **Figure 5** shows that the number of deaths due to cold wave in the Tarai region were higher in its central and eastern districts than other parts [20].

Further, the deaths of people due to cold wave in the Tarai region are found to have taken place in 4 months (**Figure 6**). The distribution of the deaths is that: it has reached peak in January, with 71.5% and followed it by December (22.0%), February (3.8%), and November (2.4%).

Figure 7 shows 2 years, 2004 and 2013, having the largest number of deaths due to cold wave in the Tarai. Though there is a fluctuation in the number of deaths due to cold wave, gradual rising of number of deaths is found since the year 2000 and more elevated trend of number of deaths since 2009.

Description	Cold wave	% of cold wave among the total impacts due to disaster
Deaths of human beings	822	2.5
Injured due to cold wave	83	0.1
Indirectly affected—morbidity	2405	0.0
Economic losses (US$)	834,650,000	2.1
Damages of crops (Ha)	26906.5	2.6
Death of cattle	732	0.1

Source: [20].

Table 5. Impacts of cold wave in Tarai region (1974–2014).

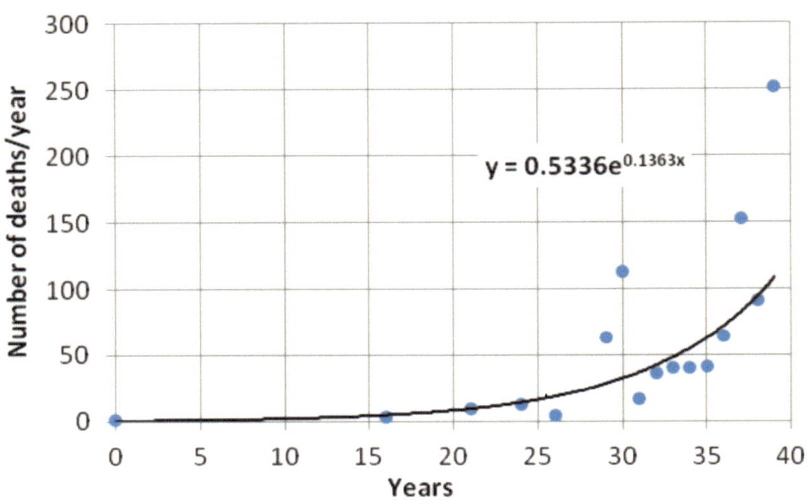

Figure 4. Death trend due to cold wave during the years 1974–2013.

3.3. Age-specific vulnerable people

The number of deaths due to cold wave is found varied remarkably among the age groups of vulnerable people. Of the total deaths, about 60% were children below the age of 5 years, while the 35% elderly population occupied 35% and others the rest 5%.

3.4. Perception of the people

3.4.1. Cause of death

The perception survey indicated that the deaths of the people in the Tarai were mainly due to severe cold, as poor people (children and elderly) with inadequate living conditions (lack of warm cloths and poor house-huts) could not combat with the impacts of severe cold wave. The deaths are found due to diseases like pneumonia, ARI, influenza, COPD, asthma, fever, and hypothermia.

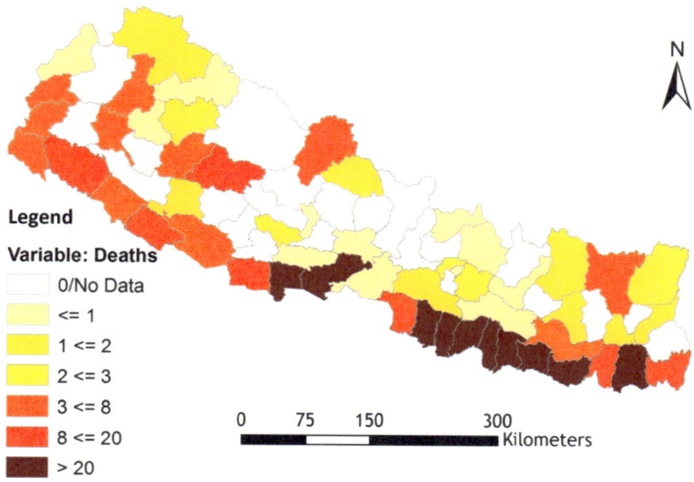

Figure 5. Spatial distribution of deaths due to cold wave by district in Nepal (1974–2013).

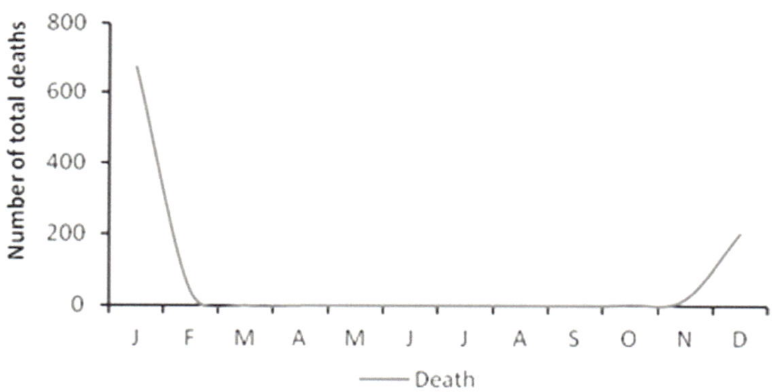

Figure 6. Total number of deaths in Tarai region from 1974 to 2013.

3.4.2. Adaptation

Undoubtedly, impact of cold wave is severe among the local communities, whose economic status is poor, and also daily wage laborers are affected the most, as their wage works are hindered due to cold wave. Normally, they use fuelwood to combat the cold wave but that is not adequate to manipulate the room temperature to bring to the normal standard. In response to this, the Government of Nepal is found distributing fire wood to the local inhabitants of certain pocket areas, and warm clothes to the new born baby and mother in the hospital during delivery.

3.5. Mortality prediction

The prediction of number of deaths of people due to cold wave has been carried out by using multiple linear regression, based on the data of three variables such as minimum temperature,

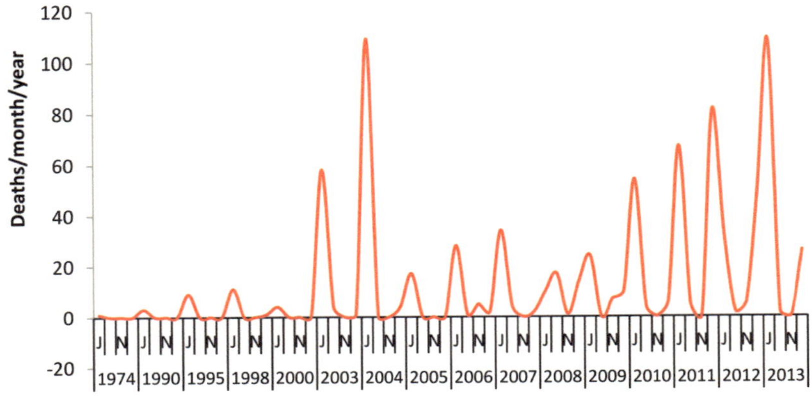

Figure 7. Mortality per month/year due to cold wave in Tarai region.

maximum temperature and rainfall from 1974 to 2014. A significant regression equation was found at [F(3, 38) = 4.258, $p < 0.05$)], with an R^2 of 0.252. The prediction of the number of deaths by employing the multiple linear regression is found equal to 814.84 + 20.07 (minimum temperature) − 40.70 (maximum temperature) − 0.561 (rainfall), where minimum temperature and maximum temperature are measured with degree centigrade (°C) and rainfall in millimeter (mm). Number of deaths of people is found to be decreased by 20 with an increase of 1° minimum temperature. On the other hand, the number of deaths is calculated at 54 people with a decrease of 1°C of temperature and 0.56 deaths with each millimeter decrease in rainfall.

The multiple regression tool has been used to predict the number of deaths from 2000 to 2013 considering the same three variables and the significant regression equation was found [F[3, 11] = 1.483, $p < 0.05$)], with an R^2 of 0.29. Prediction of the number of deaths based on the multiple linear regression is equal to 980.45 − 32.442 (minimum temperature) − 25.695 (maximum temperature) − 0.066 (ml), where temperature conditions are measured with degree centigrade (°C) and rainfall is measured in millimeter (mm). The number of deaths is calculated to increase at 32, with a decrease of each degree in minimum temperature. Similarly, 26 people are expected to die with decrease of 1° temperature and 0.066 deaths will occur with a decrease of each millimeter of rainfall.

4. Discussions

As other countries across the world, Nepal has also experienced increasing trend in average annual temperature as well as in minimum and maximum temperature conditions [26]. But however, since the last two decades, Nepal's Tarai region has got a decreasing trend of minimum temperature in winter season, unlike other regions [19]. While attempts have been made to analyze the temperature trend, two slots of year duration such as 1974–2014 and 2000–2014 were found as distinctive. In contrast to the trend of temperature conditions during

1974–2014, the analysis shows negative trends of minimum and maximum temperature conditions in the winter season during 2000–2014 years, indicating increase in the number of cold days in the Tarai. Globally, the annual numbers of warm nights and cold nights have increased and decreased by about 25 and 20 days, respectively, since 1951 [27]. In Nepal, the annual numbers of cool days and cool nights have decreased by about 5 and 9 days, respectively from 1971 to 2006, but however, during the same time period, the warm days have increased by about 16 days and warm nights have increased by only about 7 days [28]. It is interesting to note here that the decreasing trend of cool days and increasing trend of warm days are clearly seen at higher elevations in Nepal [28].

Death of people due to cold wave is increasing in the Tarai due to increasing duration of cold wave. Tarai region also suffers death of people from heat wave, as it is the hottest region of Nepal. However, the number of deaths due to cold wave is larger than that due to hot wave in the Tarai. The study of MoHA found that the number of deaths in the Tarai due to heat wave from May to August was 45 during the years 1974–2013, which was, however, quite low as compared to the 822 deaths due to cold wave from November to February during the same year duration [20]. The impact of heat wave was mostly on the elderly people, while that of cold wave was on children.

Death of people in the Tarai is found not only due to cold wave but also because of lack of facilities in living places or public hospitals [19, 26, 29]. For instance, during the severe cold months, the average indoor room temperature was found so low than the normal standard [13, 15]. Even in the hospitals, there was seldom provision of room heating system, resulted into death of neonatal and under-five children due to hypothermia and acute respiratory problem [15]. Further, there has been an increasing trend of mortality and morbidity due to respiratory diseases like ARI, COPD and cardiovascular diseases as a result of decreased temperature or cold wave in the Tarai region of Nepal [16]. The same study has predicted that there will be decrease in number of death of people by 2.68% due to ARI as per 1°C rise in minimum temperature. A study carried out in Sarlahi district of Tarai region found out that about 92% of new born babies born in winter suffered from the hypothermia [30]. Overall, only 10.7% of neonates have received optimum thermal care as per the WHO guidelines [15].

In addition to the deaths of people, there are other adverse impacts of cold wave in the Tarai region. Running schools and daily life and livelihoods of people usually are severely interrupted by the cold waves, especially, the vulnerable people like children, elderly and pregnant with low-income groups, and homeless people and daily waged people are affected the most. It is found that cold wave is a risk factor for diseases like respiratory, cardiovascular, viral influenza and Rotavirus infection [16]. Further, during the onset of cold wave, there would be poor visibility leading to increasing trend of road injuries and interruptions in aviation industry, which ultimately hinder livelihood. Outbreaks of avian influenza have a highly seasonal pattern, with nearly all outbreaks occurring in January and February [31]. In mid-winter, especially, the Tarai region can experience cold waves, which often cause crop damage that may lead to famine [20, 32].

Undoubtedly, impact of cold wave is severe on the community, whose economic status is poor, and daily wage laborers are also affected the most, as their activities are hindered due to cold wave. As they are poor, they burn locally available fuelwood or straw, which are also available in small quantity, to combat the cold wave, but this is found not adequate to manipulate the room temperature to bring to the normal standard. Government's effort in this case is very crucial. The government agencies used to distribute fire wood in some pocket areas and warm clothes to the new born baby and delivered mother in the hospitals. Until now, these mere efforts so far undertaken by the government to address the adverse impacts of cold wave seem inadequate. As health is directly and indirectly affected by climate change via various pathways, there should be a priority focus on health in national adaptation sustainable plans for the medium- and long-term needs of all sectors, such as multisectoral preparedness plans [25, 26].

5. Conclusion and recommendation

It is evident that the average minimum temperature trend during the winter months is declining in the Tarai region of Nepal. The predictions of minimum and maximum temperature trend with regard to number of deaths have been made with different scenarios, that is, increasing or decreasing of 1°C affecting the change in number of death of people by using the modest model of multiple linear regression. Number of deaths due to cold wave in the Tarai has increased over the past two decades due to increasing duration of cold waves in the winter months. Adverse impacts are seen more on vulnerable groups of population such as under-five children and elderly. These are no doubt, the impacts due to global or regional warming, change in land uses, rapid urbanization, etc. It is urgently essential that the adaptation strategy and plans should be designed and implemented to address the increasing trend of cold wave in the Tarai region and other regions based on the findings and recommendations of the rigorous studies.

Author details

Bandana Pradhan[1*], Puspa Sharma[2] and Pushkar K. Pradhan[2]

*Address all correspondence to: bandana@reachpuba.org

1 Institute of Medicine, Tribhuvan University, Kathmandu, Nepal

2 Central Department of Geography, Tribhuvan University, Kirtipur, Nepal

References

[1] Pachauri RK, Allen MR, Barros VR, Broome J, Cramer W, Christ R, et al. Climate change 2014: Synthesis report. In: Contribution of Working Groups I, II and III to the Fifth Assessment Report of the Intergovernmental Panel on Climate Change. IPCC; 2014

[2] Pachauri RK, Reisinger A. IPCC Fourth Assessment Report. Geneva: IPCC; 2007

[3] Analitis A, Katsouyanni K, Biggeri A, Baccini M, Forsberg B, Bisanti L, et al. Effects of cold weather on mortality: Results from 15 European cities within the PHEWE project. American Journal of Epidemiology. 2008;**168**(12):1397-1408

[4] Anderson GB, Dominici F, Wang Y, McCormack MC, Bell ML, Peng RD. Heat-related emergency hospitalizations for respiratory diseases in the medicare population. American Journal of Respiratory and Critical Care Medicine. 2013;**187**(10):1098-1103

[5] Arbuthnott K, Hajat S, Heaviside C, Vardoulakis S. Changes in population susceptibility to heat and cold over time: Assessing adaptation to climate change. Environmental Health: A Global Access Science Source. 2016;**15**(Suppl):1-33

[6] Ebi KL, Mills D. Winter mortality in a warming climate: A reassessment. Wiley Interdisciplinary Reviews: Climate Change. 2013;**4**(3):203-212

[7] Kysely J, Pokorna L, Kyncl J, Kriz B. Excess cardiovascular mortality associated with cold spells in the Czech Republic. BMC Public Health. 2009;**9**(1):19

[8] Cagle A, Hubbard R. Cold-related cardiac mortality in King County, Washington, USA 1980-2001. Annals of Human Biology. 2005;**32**(4):525-537

[9] Healy JD. Excess winter mortality in Europe: A cross country analysis identifying key risk factors. Journal of Epidemiology and Community Health. 2003;**57**(10):784-789

[10] Liu C, Yavar Z, Sun Q. Cardiovascular response to thermoregulatory challenges. The American Journal of Physiology—Heart and Circulatory Physiology. 2015;**309**(11): H1793-H1812

[11] Zhou MG, Wang LJ, Liu T, Zhang YH, Lin HL, Luo Y, et al. Health impact of the 2008 cold spell on mortality in subtropical China: The climate and health impact national assessment study (CHINAs). Environmental Health. 2014;**13**(1):60

[12] Malla G. Climate change and its impact on Nepalese agriculture. Journal of Agriculture and Environment. 2008;**9**:62-71

[13] Ellis M, Manandhar N, Shakya U, Manandhar D, Fawdry A, Costello A. Postnatal hypothermia and cold stress among newborn infants in Nepal monitored by continuous ambulatory recording. Archives of Disease in Childhood. Fetal and Neonatal Edition. 1996;**75**(1):F42-FF5

[14] Mullany LC, Katz J, Khatry SK, LeClerq SC, Darmstadt GL, Tielsch JM. Risk of mortality associated with neonatal hypothermia in southern Nepal. Archives of Pediatrics & Adolescent Medicine. 2010;**164**(7):650-656

[15] Khanal V, Gavidia T, Adhikari M, Mishra SR, Karkee R. Poor thermal care practices among home births in Nepal: Further analysis of Nepal demographic and health survey 2011. PLoS One. 2014;**9**(2):e89950

[16] WHO. Report on Assessment of Health Effects of Cold Waves in Terai Nepal. Kathmandu: The Green Move Consultancy; 2017

[17] CBS. Population Monograph of Nepal 2014. Government of Nepal; 2014

[18] MoH. Nepal Demographic and Health Survey 2016: Key Indicators. Kathmandu, Nepal: Ministry of Health, Nepal; New ERA; and ICF; 2016

[19] DHM. In: Meteriology DoHa, editor. Observed Climate Trend Analysis in the Districts and Physiographic Regions of Nepal (1971-2014). Kathmandu: MoPE; 2017

[20] MoHA. Nepal Disaster Report 2015. Government of Nepal, Ministry of Home Affairs (MoHA) & Disaster Preparedness Network-Nepal (DPNet-Nepal); 2015

[21] DoHS. In: Servises DoH, editor. Annual report 2013/14. Kathmandu: Ministry of Health and Population; 2015

[22] UNISDR. Disaster Database Sendai Framework Nepal [Internet]. Available from: http://www.desinventar.net/DesInventar/profiletab.jsp?countrycode=np

[23] Bolin JH, Hayes AF. Introduction to mediation, moderation, and conditional process analysis: A regression-based approach; 2013. New York, NY: The Guilford Press. Journal of Educational Measurement. 2014;**51**(3):335-337

[24] MoPE. Climate Change and Glacial Lake Outburst Floods in Nepal, Kathmandu. Kathmandu: ICEM—International Centre Environmental Management with the Nepal Ministry of Population and Environment (MoPE) as part of TA–7984 NEP; 2014

[25] MoPE. Synthesis of Stocktaking Report for National Adaptation Plan (NAP) Formulation Process in Nepal. Kathmandu: Ministry of Population and Environment; 2017

[26] MoPE. Vulnerability and Risk Assessment Framework and Indicators for National Adaptation Plan (NAP) Formulation Process in Nepal. Kathmandu: Ministry of Population and Environment (MoPE); 2017

[27] Alexander L, Zhang X, Peterson T, Caesar J, Gleason B, Tank AK, et al. Global observed changes in daily climate extremes of temperature and precipitation. Journal of Geophysical Research-Atmospheres. 2006;**111**(D5):1-22

[28] Baidya SK, Shrestha ML, Sheikh MM. Trends in daily climatic extremes of temperature and precipitation in Nepal. Journal of Hydrology and Meteorology. 2008;**5**(1):38-51

[29] Dhimal M, Dhimal ML, Pote-Shrestha RR, Groneberg DA, Kuch U. Health-sector responses to address the impacts of climate change in Nepal. WHO South-East Asia Journal of Public Health. 2017;**6**(2):9

[30] Mullany LC, Darmstadt GL, Khatry SK, Katz J, LeClerq SC, Shrestha S, et al. Topical applications of chlorhexidine to the umbilical cord for prevention of omphalitis and neonatal mortality in southern Nepal: A community-based, cluster-randomised trial. The Lancet. 2006;**367**(9514):910-918

[31] World-Bank. Project Performance Assessment Report Nepal: Avian Influenza Control Project (IDA-H2680); 2013

[32] Rohwerder B. Seasonal Vulnerability and Risk Calendar in Nepal. Applied Knowledge Services. Governance Social Development Humanitarian Conflict; 2016

Chapter 10

A Survey of Disaster Risk and Resilience in Small Island States

Tauisi Minute Taupo

Additional information is available at the end of the chapter

http://dx.doi.org/10.5772/intechopen.80266

Abstract

This paper surveys the conceptual framework of disaster risk that relies on its associated components of hazard, vulnerability and exposure. How we measure these risks depends on how we define disaster risk and its components. We focus on the implication and applicability of available conceptual frameworks of disaster risk on small and low-lying islands in the Pacific. We examine some of the available measurements of these disaster risks as they are imperative to the formulation of appropriate disaster risk reduction (DRR) policies for Tuvalu. Though there are diverse views on these definitions in different disciplines, we can capitalise on their commonalities to frame disaster risk models. Here, we intend to use the findings and set a pathway for potential research and to contribute into building resilience, reducing DRR and improving responsiveness to the impact of climatic disasters in Pacific Islands.

Keywords: disaster risk, hazard, exposure, vulnerability, resilience

1. Introduction

Natural disasters such as cyclones, earthquakes, floods, tsunamis, storm surges and heat-waves have distressed the lives of the people around the world. The Asia-Pacific region is the most highly exposed to disasters in the world, with the highest overall disaster-related deaths, representing 75% of global mortality for the years between 1970 and 2011 [1]. Changes in the climate, sea level rise and the intensity of climatic disasters like tropical cyclones[1],

[1] According to [2] small countries are more vulnerable to windstorms than other countries which can lead to a decline of 3% in GDP per capita.

droughts and floods have an extremely negative impact on economies, communities, households, people and physical assets [3, 4]. Developing countries are especially vulnerable to these impacts due to their underlying limited natural endowments, economic constraints and limited adaptive capacity [5]. Small Island Developing States (SIDS) are especially vulnerable to large-scale economic and environmental disasters, whereby their geography and size make them highly exposed and vulnerable, with less capacity to respond [6].

In the Pacific region, climate-related disaster risk has been increasing in the past decades, most likely because of increased exposure of people and economic assets. Within the Pacific SIDS, the smaller island states[2] like of Tuvalu and Kiribati consist of low-lying[3] stretches of atoll islands that are most vulnerable countries to climate change, sea level rise and climatic disasters[4], particularly to destructive cyclones with associated storm surges that can easily flood large parts of the islands. By United Nations (UN) standards, smaller island states are mostly categorised as least developed countries (LDCs). Their vulnerability, exposure and economic status slow their graduation from being LDCs. We focus on Tuvalu and Kiribati as they are low-lying atolls and sovereign states within the Smaller Pacific Island States. In Tuvalu, natural disasters such as cyclones with associated storm surges often flood some islands, inflicting significant damage on the livelihoods and physical assets of the population, while imposing adverse effects on the economy and ecosystems.

The cyclone of 1972 is the worst event ever experienced by Tuvalu.[5] However, there have been other noticeable strong storms in the recent past. A more recent event, in 2015, was a distant cyclone (about 1000 km away) called Tropical Cyclone Pam (TC Pam), affecting the islands of Tuvalu with estimated damage and losses of 10% of GDP [9]. The changes in weather patterns and the threat of rising sea levels due to climate change further aggravate these threats.

This inquiry reviews the growing body of the recent literature on disaster risk and associated components influencing it. We aim to understand the concepts of disaster risk in order to recognise its challenges, opportunities and implications for SIDS, particularly low-lying islands like Tuvalu. Through this, we can acquire ideas of what is needed to improve disaster risk management (DRM) and ways to advocate for and strengthen disaster risk reduction (DRR) efforts. Reviewing the literature on disaster risk will also situate this research in its broader context, in order to provide direction for future research in this growing field, with the focus on small island states.

[2] Smaller island states classified under the Pacific Islands Forum comprise Cook Islands, Federated States of Micronesia, Kiribati, Marshall Islands, Nauru, Niue, Palau and Tuvalu.
[3] Ref. [7] stated that the whole land in Tuvalu lies below 5 m.
[4] Vulnerable to natural disasters in per capita terms.
[5] Ref. [8] stated that 'In October 1972, cyclone "Bebe" hit Tuvalu, killing several people destroying millions of dollars worth of property. The capital atoll of Funafuti was engulfed by waves from both the ocean and lagoon side, with a huge 19 km long, 30–40 m wide and 4 m high embankment (called a "storm ridge") being formed as a consequence of the waves moving huge quantities of sediments. The storm damaged houses, infrastructure, boats, coconut trees, the reef flats and caused extensive scouring of the islets in the atoll.' (See http://www.janeresture.com/hurribebe/hurricanebebe2.htm for full details of the impact of cyclone Bebe on Funafuti Island, including documented stories from seven people who experienced the devastation of the event).

2. Disaster risk

Natural disasters affect people worldwide, causing losses and damages. Climate change and its influence on the frequency and intensity of natural disasters have been part of the emergence of the new branch of economic research on the economics of disasters.

Ref. [10] defined disaster as 'a serious disruption of the functioning of a community or a society involving widespread human, material, economic or environmental losses and impacts, which exceeds the ability of the affected community or society to cope using its own resources'. Here, disasters are being described in relation to exposure, vulnerability and coping mechanisms. The definition of disaster risk reflects on the meaning of disasters; disaster risk is not only the likelihood of a disastrous event but also often associated with mechanisms that inflate the impacts of such events. Particularly, disaster risk is a function of three interlinked components: hazard, exposure and vulnerability [11]. By definition, [10] refers to disaster risk as 'the potential disaster losses, in lives, health, status, livelihoods, assets and services, which could occur to a particular community or a society over some specified future time period'.

Refs. [12, 13] elaborate on the framework of the 'dual-faced' character of nature that presents a set of possible opportunities and possible hazards, emphasising that disasters are not solely driven by the natural environment but also influenced by human activities, that is, they are the product of political, social and economic environments. They also introduced a conceptual framework that defines and explains the relationship between risks, hazards and vulnerability. This pressure and release (PAR) framework illustrates that the intersection of hazard, vulnerability and coping and recovering capacities correspond to disaster risk. Moreover, [13] advanced the framework of 'progression and vulnerability' comprising of root causes, dynamic pressures and fragile livelihoods and unsafe locations.[6] This framework reflects the fact that limited access to resources that allow for risk reduction impedes coping and recovery mechanisms for hazards. Nevertheless, disaster risk and its underlying components (hazard, exposure and vulnerability) are changing in relation to the changes in the environment and political, economic and social aspects of society [11].

2.1. Hazard

Hazard is widely recognised as an extreme natural event or process [13], a potential harmful event or process [4], or a hazardous phenomenon [14]. In the past, natural hazards and their characteristics were the main focus of discourses relating to disasters.

In addition to naturally occurring hazards, the evolution of the way we look at disasters has unfolded new components of disaster risk and extended its scope. [10] refers to hazard as 'a dangerous phenomenon, substance, human activity or condition that may cause loss of life,

[6] 'Root causes' in this context centres around existing social, economic (e.g. distribution of resources, wealth and power) and political structures. 'Dynamic pressures' concerns with societal deficiencies (relating to economic opportunities) and lack of macro forces. Unsafe conditions are specifically associated with the situation facing vulnerable people in a given time and place.

injury or other health impacts, property damage, loss of livelihoods and services, social and economic disruption or environmental damage'. Recently, [11] defined hazard as the likelihood and intensity of a potentially destructive natural phenomenon, such as ground shaking induced by an earthquake or extreme winds associated with a cyclone.

Generally, hazard is interpreted as an influence that can adversely affect a system's valued attributes. Ref. [15] noted that there has been a paradigm shift in the development literature on hazards and disasters, from assessing hazard to analysing vulnerability and building community resilience.

2.2. Exposure

Ref. [11] defined exposure as the location, attributes and value of people and assets (such as buildings, agricultural land and infrastructure) exposed to the hazard. Ref. [4] broadly refers to exposure as the external environment that determines the shocks to which a system is subject. Ref. [16] postulated that exposure is the 'degree, duration, and/or extent in which the system is in contact with, or subject to, the perturbation'. Ref. [17] described exposure more discretely by referring to it as 'the likelihood that an individual in a given location is exposed to a given type of climate-related hazard event over a certain period of time'. They also estimated population exposure to climate-related hazards (e.g. cyclones, droughts and floods) using gridded datasets, with which they calculated the population exposure by the relative hazard frequency in a certain area weighted by the population density frequency. As a result, they ranked countries by population exposure to these extreme events.

Ref. [14] refers to exposure as the location of people, production, infrastructure, housing and other tangible human assets in hazard-prone areas. Ref. [10] defined exposure as 'people, property, systems, or other elements present hazard zones that are thereby subject to potential losses'. The poor are exposed to disasters [18], and, further, poor people are often, but not always, more exposed to hazards [19]. Ref. [20] developed an exposure model for hazard risk assessment from a Country Disaster Risk Profile (CDRP) which complements vulnerability and hazard models.

2.3. Vulnerability

Ref. [21] posited that the scientific use of 'vulnerability' has its roots in geography and natural hazard research but has become a central concept in many other research contexts. Vulnerability is defined as the potential extent to which physical, social, economic and environmental assets may become damaged or disrupted when exposed to a hazard event [11]. Vulnerability is a complex term with no consensus on its meaning, though it tends to include various factors that have the potential to be damaged or harmed by a hazard event. For instance, on a physical scale, it refers to physical vulnerability when looking at the level of damage sustained by built structures due to a hazard event. On the social level, it refers to 'social vulnerability' (also known as 'socio-economic vulnerability' or 'socio-economic resilience') where damage relates to livelihood and other social factors that influence a community's ability to respond to, cope with and recover from a disaster [11]. Social vulnerability can affect the number of casualties, the loss or disruption sustained and a community's subsequent recovery time. Similarly, [10] defined vulnerability as 'the characteristics and circumstances of a community, system or asset that make it susceptible to the damaging effects of a hazard'.

Vulnerability is often seen as a potential for weakening the capacity of an individual or group to face, cope, resist, respond to and recover from the impacts of natural and anthropogenic hazards. Ref. [22] comprehensively outlined the analysed vulnerability framework and its components—exposure, sensitivity and resilience—which are linked to dynamic factors beyond the system. The degree of vulnerability and capacity to respond to and recover from disasters are often determined by physical, economic, social and political factors. Through this, vulnerability is often connected to poverty. To better respond to these disasters, there is a need for reducing hazard impact (through preparedness, mitigation, prediction, etc.), building and strengthening capacities to resist and cope with hazards and attempting to reduce sources of vulnerability (e.g. through poverty reduction, good governance, equality, accessibility to resources and livelihoods).

Ref. [23] argued that the extent of disaster risk depends on natural hazards and vulnerability. They also support the argument that there is a higher level of vulnerability in the urban areas due to higher population density. As poorer households tend to reside in riskier urban areas [24], they are more likely to face rising costs and relatively higher losses and damages during disasters. Normally, the poor are more affected due to economic and social attributes [12]. At the macro-level, [23] further illustrates the notion of the 'inverted U' relationship between economic development and disaster vulnerability, indicating that middle-income countries are specifically vulnerable to natural disasters. Ref. [25] strengthened the link between poverty and disaster for Bangladesh arguing that the poor are not only more vulnerable to natural events but also have less ability to access resources due to factors such as social and political identity, kinship, social networks, financial capacity, political connections and rivalry. They argued that the dynamics of livelihoods, local power, resilience and cyclones are interconnected.

Ref. [26] examined climate justice for SIDS like the Caribbean islands and argued that factors driving vulnerability pointed to centuries of economic neglect and political marginalisation that are strongly related to communities' socio-economic characteristics, geographic locations, heavy reliance on land-based resources and the capacity to adapt to climate change. Refs. [26, 27] stressed that vulnerability to negative impacts of climate change is partly a function of different coping and adapting capabilities of various groups of people in developing countries. Ref. [26] further argued that vulnerability to climatic impacts is inherently developmental as differentiated levels of exposure and sensitivity to natural hazards are partly created by social and economic inequalities, as well as accessibility of land-based resources, assets and government support. Ref. [27] strengthened the notion that vulnerability and capacities to cope with natural hazards differ due to differential accessibility to resources (e.g. natural, physical, human, social and political).

On the other hand, economic vulnerability is well documented in the literature from both conceptual and empirical viewpoints. Most studies in this stream point to the small island states as highly vulnerable to exogenous shocks due to their high degrees of economic openness and export concentration [28]. SIDS are most vulnerable to disaster risks due to increasing intensity of cyclones and sea level rise. Ref. [29] recognised the vulnerability of SIDS to disasters and the lack of economic resilience arising from the relative inability of these countries to face forces of scales out of their capacity to deal with independently. Ref. [30] emphasised that measuring risks and vulnerability is imperative in promoting disaster resilience in hazard-prone areas.

2.4. Resilience

Refs. [16, 31–33] all believed that the literature on resilience emerged from the ecology discipline, as an offspring of an influential paper by [34] called 'Resilience and stability of ecological systems'. [15] outlined the various definitions of community resilience. Ref. [35] argued that definitions reflect on the nature of a system (e.g. ecological, economic, social or political). For instance, the literature and different organisations have their own definitions for disaster resilience. Ref. [33] proposed a framework called the Disaster Resilience of Place (DROP) model and emphasised that there is more to articulate about the relationship between vulnerability, resilience and adaptive capacity. They distinguish 'vulnerability' as the characteristic that creates the potential for harm and 'resilience' as the ability to respond to and recover from disasters. Similarly, although [28] argued that risk is determined by the two elements, namely, exposure and coping ability, they associate exposure to vulnerability and coping ability to resilience.

Many terms were used to describe the various efforts to reduce risk, namely, preparedness, public awareness, prevention, adaptation, resistance, mitigation, response and so on. Ref. [10] refers to resilience as 'the ability of a system, community or society exposed to hazards to resist, absorb, accommodate to and recover from the effects of a hazard in timely and efficient manner, including through the preservation and restoration of its essential basic structures and functions'. With the increasing global threat of climate change and associated disasters, measuring resilience has increased in popularity with efforts attempting to build resilience to climate change and disasters. Consequently, many papers have emerged with definitions, concepts and indicators for measuring resilience. These concepts are often complexly interrelated. Other studies by [36] identified economic indicators of resilience based on impact, outcome, output and input.

Furthermore, [37, 38] compared 'economic vulnerability' and 'economic resilience' to explain the phenomenon known as the 'Singapore Paradox'.[7] This is a phenomenon based on the fact that even though the small island state of Singapore is highly exposed to exogenous shocks, still they achieved and attained high levels of economic growth and GDP per capita.[8] The importance of economic resilience to disasters was highlighted by [36] as an enabler of many broader development goals.

Building resilience requires a clear concept of 'resilience' itself. Reflecting on the complications of civil society and the thinking behind disasters in relation to society, it is not surprising that various disciplines have diverse definitions of 'resilience'. However, there are commonalities apparent in these definitions, and this is fundamental in establishing a resilience paradigm. With more clarity and consensus on the definitions, disaster resilience can be more achievable with less confusion. Consequently, an evolution in defining 'resilience' has steadily enhanced the way we conceptualise disaster resilience.

[7] See [39] for more discussion on 'economic vulnerability' and 'economic resilience'.

[8] This case can be explained by its ability to face external shocks through building resilience. Unlike other isolated small island states like Tuvalu and Kiribati, Singapore has geographical advantages (port, location is heavily populated in the region) as a trade centre, with the presence of multinational companies on its shores. Therefore, Singapore has the potential to build economic resilience.

On a more practical level, [40] employed a resilience index measured as a ratio of preparedness (capacity to overcome a disaster) to vulnerability (exposure towards a disaster). Ref. [18] revealed the 3D resilience framework, whereby resilience arises as a result of the three capacities, namely, absorptive, adaptive and transformative capacities. On the other hand, [41] modelled infrastructure resilience by quantifying resilience as a function of absorptive, adaptive and restorative capacities. They refer 'absorptive' as the capacity of a system to absorb or withstand the impact of disruptive events and minimise the consequences, 'adaptive' as the capacity of a system to adapt and overcome a disruption, and 'restorative' as the capacity of a system to repair and restore from a disruption.

3. Measurements of disaster risk

In the economic community, there are available econometric methods to measure disaster risk (or disaster impact). Most use time series, cross section and panel data to identify the relationship of explanatory variables on disaster risk (or disaster impact). Ref. [42] showed how the impacts of disasters can be measured by referring to a model of the form: $Y_{it} = \alpha + \beta X_{it} + \gamma DIS_{it} + \varepsilon_{it}$ where Y_{it} denotes disaster impact of interest, DIS_{it} is a measure of the immediate impact of disaster on country i at time t, X_{it} is the typical vector of control variables affecting Y_{it}, and $Y_{i,t-1}$, and ε_{it} is the error term. They also show other extended models using other estimation methodologies.

Other studies like [43, 44] suggested that disaster risk is a function of hazard, vulnerability and capacity. However, discourse surrounding the definition of 'resilience' compared to 'capacity' has resulted in diverse perspectives and formulations of disaster risk. Following definitions from [10, 44] modified the formula proposed by [43] arguing that disaster risk should be a function of natural hazards, vulnerability, exposure and resilience. They argued that this modification better reflects on the underlying purpose of disaster risk reduction (DRR) in reducing vulnerability and exposure to hazard while building resilience for potential impacts. Measuring of disaster risk in this manner has become popular with the increasing intensity of disasters and associated costs (loss and damage) over time, which has in turn generated the emerging focus of research in this area followed by extensions in definitions and concepts. Ref. [45] breaks down the effects of disasters into direct damages (affected assets) and indirect losses (affected flow of goods and services).[9] These developments, extensions and interactions (capturing resilience, adaptability, responses and other factors) are useful in identifying areas in need of building resilience for disasters. This has made measuring and building of resilience an essential tool for reducing disaster risk.

The conceptual framework for risk under the PAR framework [12] discussed earlier outlines an equation of the form where $Risk = Hazard \times Vulnerability$. Nevertheless, an extension was formulated based on the [46] expressing the crucial role of natural hazards, exposure and vulnerability in measuring disaster risk. This widely used formulation within the disaster risk

[9] See [42] for more discussion on natural disasters and the economy.

community is simply expressed as $Risk = Hazard \times Exposure \times Vulnerability$. They translate disaster risk as the multiplicative function of hazard, vulnerability and exposure. Typically, it is expressed as $Y_i = \alpha_0 + \beta_1 Hazard_i + \beta_2 Exposure_i + \beta_3 Vulnerability_i + \varepsilon_i$ for cross-sectional data. On the other hand, it can be extended to $Y_{it} = \alpha_0 + \beta_1 Hazard_{it} + \beta_2 Exposure_{it} + \beta_3 Vulnerability_{it} + \varepsilon_{it}$ for panel data, where the dependent variable Y_{it} is typically represented by direct impact on either people (losses or lost lives) or assets (e.g. direct damage costs) as a result of a disaster on unit i at time t, while the explanatory variables of $Hazard_{it}$, $Exposure_{it}$ and $Vulnerability_{it}$ are vectors of characteristics and measures that represent them. Recently, [47] extended the risk framework to capture both risks to assets and well-being by including a measure of socio-economic resilience in his risk assessment.

4. Policy implications

Ref. [5] stressed that people previously affected by natural disasters are more risk averse than those unaffected. For small and low-lying islands like Tuvalu, preparedness for tsunamis is a complex problem as there are no high grounds worthy to be safe zones. The option of resettling or moving to safer places is almost impossible in Tuvalu, because of limited lands and economic and legal constraints. Most of the people live close to the coasts, not by choice but by the limited lands available.[10] For Tuvalu, it is much safer in the outer islands than in the capital island Funafuti (urban) since the elevation is a little higher, there is more land area, and they are less populated. But, most economic opportunities are in Funafuti.

Transferring financial risk through insurance does not exist in Tuvalu, but this is recognised as a financial resilience tool to extreme climatic events like cyclones [48]. Ref. [49] points out the importance of having better institutions in a country in lowering human and economic losses from natural disasters. He also postulates on the non-linear relationship between economic development and economic disaster losses. Social networks at the local level are vital for community resilience and recovery from disasters [50–52]. Often, community resilience is a foremost response to disaster impacts [53] and also acts as informal insurance after disasters [54]. For small islands like Tuvalu, where almost everyone knows their neighbours (relatives and friends) and people on their islands, local social networks and communities are central to disaster response and recovery efforts.

It is commonly agreed that climate change will displace millions of people worldwide. However, low-lying islands in the Asia-Pacific are at the forefront of both disasters and environmental change. Migration is the last option when security and the lives of the people are at high risk. Despite this global problem, there are no provisions under international laws to protect those who will be forced to migrate due to environmental causes. Ref. [55] distinguish 'economic migration' from 'distress migration' based on household resources, capabilities and decisions. 'Migration with dignity' is a concept often advocated by some Pacific leaders [56, 57].

Migration is seen as a survival strategy for people experiencing environmental problems, but not the only available strategy [58]. Ref. [59] outlines three options: stay and do nothing

[10] With overall land area of 25 km^2 and a population of 10,000 people.

and accept the costs, stay and mitigate the changes, or leave the affected areas. In relation to ecological conditions, migration decisions are complex and linked to multiple vulnerabilities; therefore, relocation can be the only sustainable option as an adaptation strategy [60].

5. Applicability of the disaster risk model on SIDS

The components of disaster risk—hazard, exposure and vulnerability—interact to determine very high risk in most SIDS. Increases in the levels of these components will increase total risk and lead to greater damages and losses associated with disasters. Understanding and quantifying these risks in order to measure and propose risk reduction options are therefore of vital importance.

The two types of extreme events that often devastate the livelihoods of the people in SIDS are tropical cyclones and droughts. Throughout the last decade, SIDS in the Pacific has experienced some of the most severe disasters in its history. To our knowledge, these were some of the few declared disasters by the government of Tuvalu since TC Bebe in 1972.[11] Further, the intensity of disaster events has been increasing, as observed in the recent TC Pam, which caused devastation even though it was a distant cyclone to low-lying islands like Tuvalu and Kiribati. Likewise, these islands experienced one of the longest droughts in 2011. These extreme events were evaluated to be national disasters for Tuvalu that forced the government to declare states of emergencies.

For these small and low-lying islands, exposure at the household level can relate to the distance of the household from the coast and its elevation, as these are likely to be some of the determining factors of disaster risk given the size of the islands and low ground elevation. While these small and low-lying islands are considered very exposed to hazards because of their geographical setting, some islands may be more vulnerable and exposed because of combinations of other factors. For instance, apart from distance to the coast and elevation, some islands have lagoons and islets that can serve as shields during strong winds, while others have none. The width of the island is another factor: for instance, the capital island Funafuti is no more than 900 m in width, with an average of 347 m on average in land width for residential areas. The impact of a cyclone depends mostly on its distance from the islands and its trajectory, but the above are some of the extra characteristics and challenges facing small and low-lying islands that need to be included in defining exposure to disasters.

A vulnerability index for the natural environment called the Environmental Vulnerability Index (EVI) was developed by the South Pacific Applied Geoscience Commission (SOPAC) and the United Nations Environmental Programme (see [61]), whereby most of the atoll islands including Tuvalu were classified as extremely vulnerable countries (see [62]). Most low-lying Pacific Islands are least developed country (LDC), and as such, they have always been vulnerable to climatic disasters, mainly due to their geographical settings and economic

[11] TC Bebe track passed over Tuvalu, specifically between the islands of Funafuti and Nukulaelae, based on the tropical cyclone map by the Australian Severe Weather at http://www.australiasevereweather.com/tropical_cyclones/1972_1973/jtwc/tropical_cyclone_bebe.htm.

characteristics. Population density on the capital islands, for instance, Funafuti of Tuvalu where the population density is 2220 people per square kilometre, is another issue.[12] The constant increasing population density in hazardous areas [49] is evident on Funafuti with the population distribution of 42.5% (in 1991), 47% (in 2002) and 57.2% (in 2012).[13] However, socio-economic characteristics such as access to resources, communications and transportation (particularly for those in the outer islands) make some people more vulnerable than others.

Ref. [64] measured the burden of disasters on PICs using two global datasets [Emergency Events Database (EMDAT) and the Disaster Inventory System website (desinventar.net)] and concluded that these commonly used datasets immensely underestimate the burden of disasters on the PICs, particularly atoll nations [65]. He also compared the burden of disasters between PICs and the Caribbean islands to find that the burden of disasters is far more significant in the Pacific and also identified Tuvalu as the most exposed country in per capita terms. In per capita terms, the Pacific Islands face the highest disaster risk globally [65].

With increasing occurrences of cyclones in the Pacific region, Tuvalu has to strengthen DRM, response and coordination efforts and reduce disaster risk (e.g. prevention, preparedness and early warning systems). Here, we need to understand first the hazards and the exposure and vulnerability of people and assets to those hazards. Being able to identify exposure and vulnerability and quantify the current risks and potential impacts of hazards is crucial in making decisions for prevention.

6. Conclusion

The conceptual framework of disaster risk discussed is an essential stepping stone for more research, thus contributing to more knowledge about disaster risk and DRM in SIDS. Firstly, there is a need to examine the vulnerability and exposure of Tuvalu to climatic disasters at the household level. As such, using the available household data, geographic and topographic information are crucial in assessing exposure differentials between households. Here, we can also construct hardship profiles, hardship and exposure maps for households and islands to determine who are more likely to reside in highly exposed areas to disasters. With geo-coded locations linked to household surveys, we can employ spatial regression models.

Secondly, to truly examine the impact of disasters on Tuvalu, we need to conduct a household survey to quantify the impacts of a disaster following the conceptual framework of disaster risk and its associated components of hazard, exposure and vulnerability. We can also extend this conceptual framework to include responsiveness to disasters as an additional component of disaster risk. Through this, we can estimate loss and damage costs and construct

[12] Author's calculations based on the Tuvalu Census 2012 in [63]. For comparison purposes, [7] stated that Tuvalu is 'relatively highly densely populated at 437 people per sq km compared with an estimated 337 for India (1997), although Bangladesh one of the most densely populated countries in the planet has an estimated density of around 883 people per sq km'.

[13] See [63]. In absolute values, population on Funafuti in 1991, 2002 and 2012 were 3839, 4492 and 6194, respectively.

hypothetical policy scenarios for disaster risk reduction policies. Given the geographical settings of low-lying atoll islands, it is imperative to assess the impact of distant cyclones and its associated storm surge that often lead to flooding in the islands.

Last but not least, SIDS often face financial difficulties imposed by climate and disaster risks, especially for quick response and recovery. The fact that some SIDS do not have an insurance mechanism and often rely on aid for disasters stimulates our interest in developing a potential financial instrument for disaster risk management. Although there are other potential financial instruments, one option is to use their sovereign wealth funds (SWF) to contribute into a disaster fund that can then be used as a buffer for ex-post disaster risk management. We can quantify appropriate financial levels of support for expected disasters by calculating expected average annual loss (AAL). Moreover, we can assess the long-term sustainability of the SWF by forecasting its expected performance and therefore determine the feasibility of contributing to a disaster fund.

Author details

Tauisi Minute Taupo

Address all correspondence to: tauisi.taupo@usp.ac.fj

University of the South Pacific, Suva, Fiji

References

[1] UNESCAP, UNISDR. Reducing Vulnerability and Exposure to Disasters: The Asia Pacific Disaster Report 2012. Thailand: United Nations; 2012. Retrieved from: http://www.unisdr.org/files/29288_apdr2012finallowres.pdf

[2] Raddatz C. The Wrath of God: Macroeconomic Costs of Natural Disasters. Washington DC: World Bank; 2009. Retrieved from: https://elibrary.worldbank.org/doi/pdf/10.1596/1813-9450-5039

[3] World Bank & United Nations. Natural Hazards, UnNatural Disasters: The Economics of Effective Prevention. Washington DC: The World Bank; 2010. Retrieved from: http://elibrary.worldbank.org/doi/book/10.1596/978-0-8213-8050-5

[4] World Bank. World Development Report 2014: Risk and Opportunity–Managing Risk for Development. Washington DC: The World Bank; 2013. Retrieved from: http://elibrary.worldbank.org/doi/book/10.1596/978-0-8213-9903-3

[5] Ahsan DA. Does natural disaster influence people's risk preference and trust? An experiment from cyclone prone coast of Bangladesh. International Journal of Disaster Risk Reduction. 2014;**9**:48-57. DOI: 10.1016/j.ijdrr.2014.02.005

[6] Barnett J, Waters E. Rethinking the vulnerability of small Island States: Climate change and development in the Pacific Islands. In: The Palgrave Handbook of International Development. London: Palgrave Macmillan; 2016. pp. 731-748. DOI: 10.1057/978-1-137-42724-3_40

[7] McLeod H. Tuvalu Environmental Vulnerability Index Data Gathering Report. SOPAC; 1999. Retrieved from: http://ict.sopac.org/VirLib/PR0264.pdf

[8] Yamamoto L, Esteban M. Atoll Island States and International Law. Berlin, Heidelberg: Springer; 2014. DOI: 10.1007/978-3-642-38186-7

[9] ADB. Pacific Economic Monitor Midyear Review July. Asian Development Bank; 2015. Retrieved from: http://reliefweb.int/sites/reliefweb.int/files/resources/Pacific%20Economic%20Monitor_July%202015%20Midyear%20Review.pdf

[10] UNISDR. 2009 UNISDR Terminology on Disaster Risk Reduction. United Nations International Strategy for Disaster Reduction; 2009. Retrieved from: http://www.unisdr.org/files/7817_UNISDRTerminologyEnglish.pdf

[11] GFDRR. The Making of a Riskier Future: How our Decisions are Shaping Future Disaster Risk. Global Facility for Disaster Reduction and Recovery; 2016. Retrieved from: https://www.gfdrr.org/sites/default/files/publication/Riskier%20Future.pdf

[12] Wisner B, Blaikie P, Cannon T, Davis I. At Risk: Natural Hazards, People's Vulnerability and Disasters. 2nd ed. New York: Routledge; 2003

[13] Wisner B, Gaillard JC, Kelman I. Framing disaster: Theories and stories seeking to understand hazards, vulnerability and risk. In: Handbook of Hazards and Disaster Risk Reduction. London: Routledge; 2012. pp. 18-33. Retrieved from: https://www.routledgehandbooks.com/doi/10.4324/9780203844236.ch3

[14] United Nations. 2015 Global Assessment Report on Disaster Risk Reduction. New York: United Nations Office for Disaster Risk Reduction; 2015

[15] Ainuddin S, Routray JK. Community resilience framework for an earthquake prone area in Baluchistan. International Journal of Disaster Risk Reduction. 2012;**2**:25-36. DOI: 10.1016/j.ijdrr.2012.07.003

[16] Gallopín GC. Linkages between vulnerability, resilience, and adaptive capacity. Global Environmental Change. 2006;**16**(3):293-303. DOI: 10.1016/j.gloenvcha.2006.02.004

[17] Christenson E, Elliott M, Banerjee O, Hamrick L, Bartram J. Climate-related hazards: A method for global assessment of urban and rural population exposure to cyclones, droughts, and floods. International Journal of Environmental Research and Public Health. 2014;**11**(2):2169-2192. DOI: 10.3390/ijerph110202169

[18] Béné C, Wood RG, Newsham A, Davies M. Resilience: New Utopia or New Tyranny? Reflection about the potentials and limits of the concept of resilience in relation to vulnerability reduction programmes. IDS Working Papers. 2012;**2012**(405):1-61. DOI: 10.1111/j.2040-0209.2012.00405.x

[19] Hallegatte S, Bangalore M, Bonzanigo L, Fay M, Kane T, Narloch U, et al. Shock Waves: Managing the Impacts of Climate Change on Poverty. The World Bank; 2015. Retrieved from: http://elibrary.worldbank.org/doi/book/10.1596/978-1-4648-0673-5

[20] Gunasekera R, Ishizawa O, Aubrecht C, Blankespoor B, Murray S, Pomonis A, et al. Developing an adaptive global exposure model to support the generation of country disaster risk profiles. Earth-Science Reviews. 2015;**150**:594-608. DOI: 10.1016/j.earscirev.2015.08.012

[21] Füssel H-M. Vulnerability: A generally applicable conceptual framework for climate change research. Global Environmental Change. 2007;**17**(2):155-167. DOI: 10.1016/j.gloenvcha.2006.05.002

[22] Turner BL, Kasperson RE, Matson PA, McCarthy JJ, Corell RW, Christensen L, et al. A framework for vulnerability analysis in sustainability science. Proceedings of the National Academy of Sciences of the United States of America. 2003;**100**(14):8074-8079. DOI: 10.1073/pnas.1231335100

[23] De Haen H, Hemrich G. The economics of natural disasters: Implications and challenges for food security. Agricultural Economics. 2007;**37**(s1):31-45

[24] Adger WN. Vulnerability. Global Environmental Change. 2006;**16**(3):268-281. DOI: 10.1016/j.gloenvcha.2006.02.006

[25] Nadiruzzaman M, Wrathall D. Participatory exclusion – Cyclone Sidr and its aftermath. Geoforum. 2015;**64**:196-204. DOI: 10.1016/j.geoforum.2015.06.026

[26] Smith R-AJ, Rhiney K. Climate (in)justice, vulnerability and livelihoods in the Caribbean: The case of the indigenous Caribs in northeastern St. Vincent. Geoforum. 2015. DOI: 10.1016/j.geoforum.2015.11.008

[27] López-Marrero T, Wisner B. Not in the same boat: Disasters and differential vulnerability in the insular Caribbean. Caribbean Studies. 2012;**40**(2):129-168

[28] Briguglio L, Cordina G, Farrugia N, Vella S. Economic vulnerability and resilience: Concepts and measurements. Oxford Development Studies. 2009;**37**(3):229-247. DOI: 10.1080/13600810903089893

[29] Briguglio L. Small island developing states and their economic vulnerabilities. World Development. 1995;**23**(9):1615-1632

[30] Asadzadeh A, Kötter T, Zebardast E. An augmented approach for measurement of disaster resilience using connective factor analysis and analytic network process (F'ANP) model. International Journal of Disaster Risk Reduction. 2015;**14**(Part 4):504-518. DOI: 10.1016/j.ijdrr.2015.10.002

[31] Folke C. Resilience: The emergence of a perspective for social–ecological systems analyses. Global Environmental Change. 2006;**16**(3):253-267. DOI: 10.1016/j.gloenvcha.2006.04.002

[32] Mayunga JS. Understanding and applying the concept of community disaster resilience: A capital-based approach. Summer Academy for Social Vulnerability and Resilience Building. 2007;**1**:16

[33] Cutter SL, Barnes L, Berry M, Burton C, Evans E, Tate E, et al. A place-based model for understanding community resilience to natural disasters. Global Environmental Change. 2008;**18**(4):598-606. DOI: 10.1016/j.gloenvcha.2008.07.013

[34] Holling CS. Resilience and stability of ecological systems. Annual Review of Ecology and Systematics. 1973;**4**(1):1-23

[35] Smit B, Burton I, Klein RJ, Street R. The science of adaptation: A framework for assessment. Mitigation and Adaptation Strategies for Global Change. 1999;**4**(3):199-213

[36] Mitchell T, Jones L, Lovell E, Comba E. Disaster Risk Management in Post-2015 Development Goals. London: Overseas Development Institute; 2013

[37] Briguglio L. The vulnerability index and small island developing states: A review of conceptual and methodological issues. Presented at the AIMS Regional Preparatory Meeting on the Ten Year Review of the Barbados Programme of Action, Cape Verde; 2003. Retrieved from: http://www.um.edu.mt/__data/assets/pdf_file/0019/44137/vulnerability_paper_sep03.pdf

[38] Briguglio L. Economic vulnerability and resilience: Concepts and measurements. In: Economic Vulnerability and Resilience of Small States. Islands and Small States Institute of the University of Malta and Commonwealth Secretariat; 2004

[39] Noy I, Yonson R. A Survey of the Theory and Measurement of Economic Vulnerability and Resilience to Natural Hazards; 2016. Retrieved from: http://researcharchive.vuw.ac.nz/handle/10063/4978

[40] Kusumastuti RD, Viverita, Husodo ZA, Suardi L, Danarsari DN. Developing a resilience index towards natural disasters in Indonesia. International Journal of Disaster Risk Reduction. 2014;**10**(Part A):327-340. DOI: 10.1016/j.ijdrr.2014.10.007

[41] Hosseini S, Barker K. Modeling infrastructure resilience using Bayesian networks: A case study of inland waterway ports. Computers & Industrial Engineering. 2016;**93**:252-266. DOI: 10.1016/j.cie.2016.01.007

[42] Cavallo E, Noy I. Natural disasters and the economy—A survey. International Review of Environmental and Resource Economics. 2011;**5**(1):63-102. DOI: 10.1561/101.00000039

[43] Yodmani S. Disaster risk management and vulnerability reduction: Protecting the poor. Presented at the Asia and Pacific Forum on Poverty. Asian Development Bank; 2001

[44] Shaw R, Mallick F, Islam A, editors. Disaster Risk Reduction Approaches in Bangladesh. Tokyo, Japan: Springer; 2013. DOI: 10.1007/978-4-431-54252-0

[45] ECLAC. Handbook for Estimating the Socio-Economic and Environmental Effects of Disasters. United Nations, ECLAC & International Bank for Reconstruction & Development (The World Bank); 2003. Retrieved from http://eprints.mdx.ac.uk/3979/

[46] IPCC. In: Field CB, editor. Managing the Risks of Extreme Events and Disasters to Advance Climate Change Adaption: Special Report of the Intergovernmental Panel on Climate Change. New York, NY: Cambridge University Press; 2012

[47] Hallegatte S, Vogt-Schilb A, Bangalore M, Rozenberg J. Unbreakable: Building the Resilience of the Poor in the Face of Natural Disasters. Washington DC, World Bank; 2017. Retrieved from https://openknowledge.worldbank.org/handle/10986/25335

[48] Surminski S, Oramas-Dorta D. Flood insurance schemes and climate adaptation in developing countries. International Journal of Disaster Risk Reduction. 2014;**7**:154-164. DOI: 10.1016/j.ijdrr.2013.10.005

[49] Raschky PA. Institutions and the losses from natural disasters. Natural Hazards and Earth System Sciences. 2008;**8**(4):627-634

[50] Hawkins RL, Maurer K. Bonding, bridging and linking: How social capital operated in New Orleans following Hurricane Katrina. British Journal of Social Work. 2010;**40**(6):1777-1793. DOI: 10.1093/bjsw/bcp087

[51] Aldrich DP. The power of people: Social capital's role in recovery from the 1995 Kobe earthquake. Natural Hazards. 2011;**56**(3):595-611. DOI: 10.1007/s11069-010-9577-7

[52] Islam R, Walkerden G. Social networks and challenges in government disaster policies: A case study from Bangladesh. International Journal of Disaster Risk Reduction. 2017;**22**:325-334. DOI: 10.1016/j.ijdrr.2017.02.011

[53] Airriess CA, Li W, Leong KJ, Chen AC-C, Keith VM. Church-based social capital, networks and geographical scale: Katrina evacuation, relocation, and recovery in a New Orleans Vietnamese American community. Geoforum. 2008;**39**(3):1333-1346. DOI: 10.1016/j.geo forum.2007.11.003

[54] Sanyal S, Routray JK. Social capital for disaster risk reduction and management with empirical evidences from Sundarbans of India. International Journal of Disaster Risk Reduction. 2016;**19**:101-111. DOI: 10.1016/j.ijdrr.2016.08.010

[55] Johnson CA, Krishnamurthy K. Dealing with displacement: Can "social protection" facilitate long-term adaptation to climate change? Global Environmental Change. 2010;**20**(4):648-655. DOI: 10.1016/j.gloenvcha.2010.06.002

[56] Maclellan N. Kiribati's policy for "migration with dignity". 2012, January 12. Retrieved from: July 21, 2017. http://devpolicy.org/kiribati_migration_climate_change20120112/

[57] Farquhar H. Migration with dignity: Towards a New Zealand response to climate change displacement in the Pacific. Victoria University of Wellington Law Review. 2015;**46**:29

[58] Koubi V, Spilker G, Schaffer L, Bernauer T. Environmental stressors and migration: Evi dence from Vietnam. World Development. 2016;**79**:197-210. DOI: 10.1016/j.worlddev. 2015.11.016

[59] Reuveny R. Climate change-induced migration and violent conflict. Political Geography. 2007;**26**(6):656-673. DOI: 10.1016/j.polgeo.2007.05.001

[60] Marino E. The long history of environmental migration: Assessing vulnerability construction and obstacles to successful relocation in Shishmaref, Alaska. Global Environmental Change. 2012;**22**(2):374-381. DOI: 10.1016/j.gloenvcha.2011.09.016

[61] SOPAC & UNEP. Building resilience in SIDS: The environment vulnerability index. South Pacific Applied Geoscience Commission (SOPAC) and United Nations Environment Programme (UNEP); 2005. Retrieved from: http://www.vulnerabilityindex.net/wp-content/uploads/2015/05/EVI-Final-Report-2005.pdf

[62] SOPAC. EVI report: The environmental vulnerability index (EVI) 2004. South Pacific Applied Geoscience Commission; 2004. Retrieved from: http://www.vulnerabilityindex.net/wp-content/uploads/2015/05/EVI%202004%20Technical%20Report.pdf

[63] Tuvalu Government. 2012 Population and housing census: Preliminary analytical report. Tuvalu Government; 2012. Retrieved from: http://prdrse4all.spc.int/system/files/census_2012_preliminary_report.pdf

[64] Noy I. Natural disasters in the Pacific Island Countries: New measurements of impacts. Natural Hazards. 2016;**84**(S1):7-18. DOI: 10.1007/s11069-015-1957-6

[65] Noy I, Edmonds C. The economic and fiscal burdens of disasters in the Pacific. SEF Working Papers 25/2016—Victoria University of Wellington; 2016. Retrieved from: https://papers.ssrn.com/sol3/papers.cfm?abstract_id=2905652